Understanding *& Applying* *Science* 4

Sandra Baggley
Formerly GRIST Science Organiser for Manchester, now Curriculum Manager for Manchester Hospital School Service

Christopher Cammiss
Head of Physics, Whalley Range High School, Manchester

Jennifer Gow
TVEI Co-ordinator, Burnage High School, Manchester

Philip Noone
Science Teacher, Burnage High School, Manchester

Series Consultants
Joe Boyd
Walter Whitelaw

JOHN MURRAY

...otograph (courtesy of Dave Ellison, Telegraph Colour Library) shows
...emical plant at Runcorn, Cheshire

First published 1992
by John Murray (Publishers) Ltd
50 Albemarle Street, London W1X 4BD

Layout by Amanda Hawkes
Typeset by Wearset, Boldon, Tyne and Wear
Printed by Mateu Cromo Artes Graficas S.A., Madrid

A catalogue entry for this title can be obtained from the British Library

ISBN 0–7195–5040–8 Pupils' book
ISBN 0–7195–5041–6 Teachers' resource book

Contents

Extensions

Why is note making important?

Once you start to study seriously for examinations, you will find that there are lots of new facts and ideas to learn. Often you will need to use textbooks and other printed material. Making good notes helps you to learn the material and later the notes make revision very much easier.

How do you start?

You need to read the material carefully. Sometimes it is useful to **'skim'** first, to understand the general idea of what you are reading. While thinking about the general idea, read through each paragraph or section. Try to find the main point in the paragraph: sometimes there is a heading which gives you a clue. The rest of the paragraph usually gives extra information or explains the main idea more fully. When you have a clear idea of the whole passage, decide on the best way of **summarising** this information.

What should your notes be like?

Notes should be shorter than the original text! They should be brief but contain all the **important facts** and **key ideas**. Points should be arranged **logically** and follow on from each other. Try to make your notes personal by organising and presenting the information in a way that suits you. Reorganising the information mentally and putting it in your own words helps you to understand and remember it.

What layout should you choose?

This often depends on the kind of information about which you are making notes. Sometimes notes need to be written down in **sentences**. Headings, underlining and capital letters are useful for highlighting key ideas and how they are broken down into narrower points. **Flow diagrams** (spider diagrams) are useful for summarising a description of a process, a sequence of ideas or instructions. **Tables** are good for making comparisons or bringing together lots of small items of information. **Diagrams** are essential for naming different parts of an object, for describing the functions of different parts and for showing how different parts are arranged together. Use colour to make your notes more attractive and to show how different ideas or facts are related. If you are at all artistic, make drawings—they are personal and make your notes really your own.

How should you use your notes?

Once you have made your notes you should look over them fairly frequently. This fixes the information in your mind. Notes should be a 'memory jogger'. If you find that you cannot understand them when you read them again (maybe weeks later) something has gone wrong. Maybe you made them too brief or perhaps you did not give enough thought to them when you made them and there is no memory to jog!

Opposite: two examples of notes made from page 132 of this book

Ways that endotherms reduce heat loss

GENERAL WAY	WHAT IT INVOLVES	HOW IT WORKS
Insulating layer on surface of skin	Hair/fur/feathers raised	Trap more air which is a poor conductor of heat. Trapped still air also reduces heat loss by convection
	Clothes put on	
Vaso-constriction	Blood capillaries close to skin surface get narrower, reducing amount of blood flowing close to surface	Rate of heat loss falls (since skin temperature lower, and rate of conduction depends on temperature difference between surface and surroundings)
Insulating layer under surface	Layer of fat/blubber under the skin	Fat/blubber is a poor conductor of heat (good insulator)
Behavioural means	Animals seek warmer surroundings, e.g. by sheltering/burrowing/huddling	Reduces rate of heat loss to surroundings
Generating more heat	Being more active, e.g. shivering - contraction & relaxation of muscles	Generates heat which passes to the circulation
	Increasing the basal metabolic rate	Heat results from increasing rate of breakdown of food

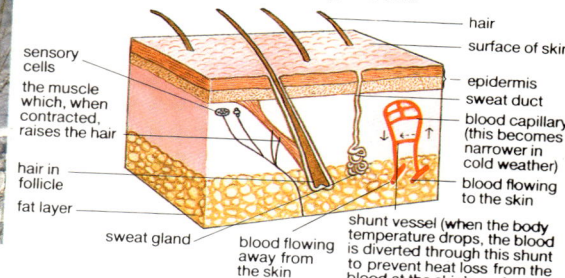

Reducing heat loss

Animals can cut down the amount of heat that they lose to their surroundings but they cannot stop heat loss completely.

The skin plays a large part in reducing heat loss.

Hairy or furry animals can raise and lower individual hairs by means of a special muscle near to the **hair follicle**. When the hairs are raised, more air can be trapped. Air is a poor conductor of heat and trapped air also cuts heat loss by convection. Radiation of heat is also reduced because the furry surface is cool. Feathers and clothes are good insulators for the same reasons. Wet fur is a poor insulator because the trapped air is replaced by water which is a much better conductor of heat.

If the blood vessels which run close to the surface of the skin become narrower, the amount of blood flowing through them is reduced. This is called **vaso-constriction** and it cuts down heat loss from the surface of the skin.

A third way that the skin controls heat loss is by having a layer of **fat** under the surface. Fat is a poor conductor of heat. All animals have some fat; those that live in a very cold environment have a much thicker layer called **blubber**.

Heat loss can be reduced by behavioural means, e.g. by going somewhere warm and sheltering. Humans shelter in buildings. Other animals, particularly small ones, dig burrows and line them with insulating material. Some animals huddle in groups. The air in the middle of the group is stiller and warmer than the surroundings.

Generating heat

One way to generate heat is by being more active. When muscles contract to produce movement they also produce heat. This heat is circulated round the body in the blood stream. **Shivering** is the constant contracting and relaxing of the muscles that the body sets off automatically as its temperature drops. Many animals also generate heat by increasing their basal metabolic rate (see p. 128).

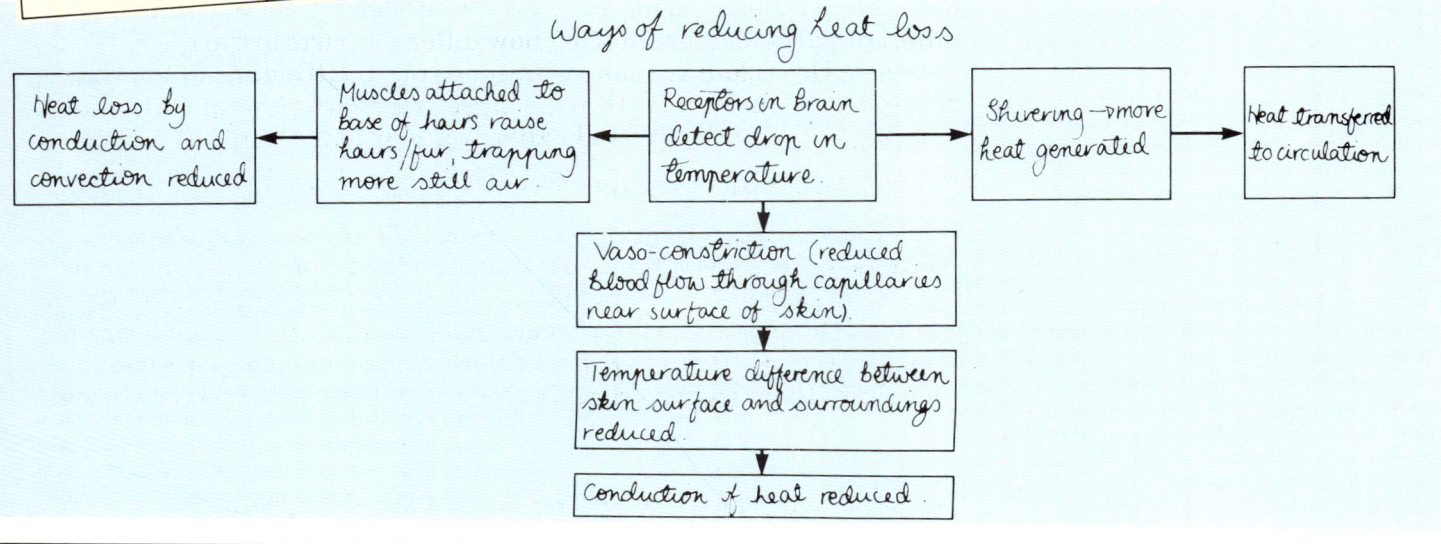

Ways of reducing heat loss

Heat loss by conduction and convection reduced ← Muscles attached to base of hairs raise hairs/fur, trapping more still air ← Receptors in Brain detect drop in temperature. → Shivering → more heat generated → Heat transferred to circulation

Receptors in Brain detect drop in temperature. ↓
Vaso-constriction (reduced blood flow through capillaries near surface of skin). ↓
Temperature difference between skin surface and surroundings reduced. ↓
Conduction of heat reduced.

Acknowledgements

CARTOONS: Ainslie MacLeod
COLOUR ARTWORK: Nancy Sutcliffe, Linda Costello
LINE DRAWINGS: Art Construction, Taurus Graphics
COVER PHOTO (and p.171): © Dave Ellison/Telegraph Colour Library

The following have provided photos or given permission for copyright photos or extracts to be reproduced:

p.9 William S. Paton/Bruce Coleman Ltd
p.10 *top* Dr B. Booth/G.S.F. Picture Library; *middle* David Woodfall/NHPA; *bottom* M.B. Withers/Frank Lane Picture Agency
p.11 Hugh Clark/Frank Lane Picture Agency
p.13 Adrian Davies/Bruce Coleman Ltd
p.14 Dr B. Booth/G.S.F. Picture Library
p.15 *graphics* Jenny Ridley/The Guardian; *photo* Jane Burton/Bruce Coleman Ltd
p.16 *maps* Ordnance Survey/© Crown copyright; *left photo* David Purdie; *right photo* M.J. Thomas/Frank Lane Picture Agency
p.18 *top* Seaphot Ltd/Planet Earth Pictures; *middle* A.N.T. (Kelvin Aitken)/NHPA; *bottom* J. Waters/G.S.F. Picture Library
p.20 *top* Jan Hinsch/Science Photo Library; *bottom* Andrew Syred/Science Photo Library
p.21 *left* Greenpeace/Grace; *right* G.S.F. Picture Library
p.23 *top photo* Michael Arron; *article* Knutsford Guardian
p.25 Jack Finch/Science Photo Library
p.26 David Purdie
p.27 Sally & Richard Greenhill
p.29 *top left, bottom right* Dr B. Booth/G.S.F. Picture Library; *top middle* Maroon/ZEFA; *top right* Royal Greenwich Observatory/Science Photo Library; *bottom left* ZEFA; *bottom middle* David Purdie
p.30 ZEFA
p.32 Robin Scagell/Science Photo Library
p.33 John McFarland/Science Photo Library
p.35 *middle left* Gerolf Kalt/ZEFA; *middle right* David Purdie; *bottom* Alfred Pasieka/Bruce Coleman Ltd
p.36 *top* L. Provo/ZEFA; *middle* Ron Boardman/Life Science Images
p.39 *top left, top right, bottom left* Ron Boardman/Life Science Images; *bottom right* C. Voigt/ZEFA
p.41 G.S.F. Picture Library
p.42 *all* Mike Williams
p.43 *top* G.S.F. Picture Library; *second top* Mike Williams; *galena* Ian Stewart Millington/Bruce Coleman Ltd; *Blue John* Jane Burton/Bruce Coleman Ltd; *malachite* A. Fisher/G.S.F. Picture Library
p.44 *all* Dr B. Booth/G.S.F. Picture Library
p.45 Sydney W. Newbery/Chatsworth
p.49 *top* Simon Fraser/Science Photo Library; *bottom* Adam Hart-Davis/Science Photo Library
p.52 Andy Purcell/Bruce Coleman Ltd
p.53 Dr F. Taylor/G.S.F. Picture Library
p.54 *a,b,d* G.S.F. Picture Library; *c* Biophoto Associates/Science Photo Library; *e* Peter Dean/Frank Lane Picture Agency
p.58 *left* Health & Safety Executive/© Crown

copyright; *right* David Woodfall/NHPA
p.59 *middle* G.S.F. Picture Library; *right* Harwood/Ecoscene
p.63 *both* David Purdie
p.65 The National Grid Company plc
p.70, p.72, p.73 Dr B. Booth/G.S.F. Picture Library
p.74 *top left* W. Mähl/ZEFA; *top right* ZEFA; *middle* (*toy, washing machine, vacuum cleaner*) Dr B. Booth/G.S.F. Picture Library; *bottom* David Parker/Science Photo Library
p.75 David Purdie
p.76 The National Grid Company plc
p.79 John Townson/Creation
p.80 BICC Cables/MPS Photography
p.81 Salim Patel/Science Photo Library
p.82 *top, bottom* David Purdie; *middle two* Dr B. Booth/G.S.F. Picture Library
p.85 J. Fennell/Bruce Coleman Ltd
p.86 *top left, top right* Dr B. Booth/G.S.F. Picture Library; *top middle* P.A. Hinchliffe/Bruce Coleman Ltd; *lower left* Aga-Rayburn; *middle* Michael McKinnon/Planet Earth Pictures
p.88 Dr. B. Booth/G.S.F. Picture Library
p.89 *top left* Steel/ZEFA; *right* J.F. Millies/ZEFA; *lower left* K & H Benser/ZEFA
p.95 Berssenbrugge/ZEFA
p.96 Gryniewicz/Ecoscene
p.97 Janulewicz/Ecoscene
p.99 *left top* Anthony Bannister/NHPA; *left middle* Michael Klinec/Bruce Coleman Ltd; *left bottom* M.P. Kahl/Bruce Coleman Ltd; *right top* L.C. Marigo/Bruce Coleman Ltd; *right bottom* Mark Boulton/Bruce Coleman Ltd
p.101 John Lee/Planet Earth Pictures
p.104 *both* Ann Ronan Picture Library (*left* from *The Illustrated Midland News*, Birmingham, 1 January 1870; *right* from *Traité de la Physique*, A. Ganot, Paris, 1862)
p.105 *top left* Ann Ronan Picture Library (from *Les Merveilles de la Science*, Louis Figuier, Paris, n.d. (c.1870)); *top right, lower left* Mary Evans Picture Library
p.106 John Lee/Planet Earth Pictures
p.107 *top left, top right* Ken Lucas/Planet Earth Pictures; *top middle* Laurie Campbell/NHPA; *bottom left* Nuclear Electric; *bottom right* U.S. Dept. of Energy/Science Photo Library
p.108 U.S. Dept. of Energy/Science Photo Library
p.110 *both* NASA/Bruce Coleman Ltd
p.116 Plaxton Coach & Bus
p.121 Mary Evans Picture Library
p.123 Transport and Road Research Laboratory/© Crown copyright
p.125 Leonard Lee Rue/Bruce Coleman Ltd
p.126 John Townson/Creation
p.127 *both* Adam Hart-Davis/Science Photo Library
p.129 *left* G.S.F. Picture Library; *right* K.H. Switak/NHPA
p.132 *top* Stephen J. Krasemann/Bruce Coleman Ltd; *bottom* Stephen J. Krasemann/NHPA
p.133 *top* Gérard Lacz/NHPA; *bottom* Anthony Bannister/NHPA
p.134 *all* Dr B. Booth/G.S.F. Picture Library

p.135 *top left* James Steveson/Science Photo Library; *right* Ron Boardman/Life Science Images; *bottom* Moredun Animal Health Ltd/Science Photo Library
p.137 *both* Ron Boardman/Life Science Images
p.140 G.S.F. Picture Library
p.144 *top* Stephen Dalton/NHPA; *bottom* Trevor J. Hill
p.146 *left* W. Wisniewski/Frank Lane Picture Agency; *right* Jill Sneesby/Bruce Coleman Ltd
p.147 *top left, top right* Colorsport; *bottom left* Billon/Colorsport; *bottom right* Tony Duffy/All Sport
p.149 Andrew Mounter/Planet Earth Pictures
p.150 *middle* Hradil/ZEFA; *bottom left* Jonathon T. Wright/Bruce Coleman Ltd; *bottom right* David Austen/Bruce Coleman Ltd
p.154 CERN/Science Photo Library
p.156 Dr B. Booth/G.S.F. Picture Library
p.157 *left* Jerry Mason/New Scientist/Science Photo Library; *right* Chris Priest/Science Photo Library
p.158 *both*, p.159 Dr B. Booth/G.S.F. Picture Library
p.160 *top left, top middle, bottom right* Dr B. Booth/G.S.F. Picture Library; *top right* Mike McNamee/Science Photo Library; *bottom left* Lucas Aerospace Ltd
p.162 Dr B. Booth/G.S.F. Picture Library
p.163 *left* Dr B. Booth/G.S.F. Picture Library; *right* A. Fisher/G.S.F. Picture Library
p.164 *left* ZEFA; *middle* G.S.F. Picture Library; *right* James Holmes/Rover/Science Photo Library
p.168 Dr B. Booth/G.S.F. Picture Library
p.173 Jan Hinsch/Science Photo Library
p.174 K. Goebel/ZEFA
p.175 P. Wilkinson/Bruce Coleman Ltd
p.177 Camera Hawaii/ZEFA
p.180, p.184 David Purdie
p.185 The National Grid Company plc
p.186 Lucas Automotive
p.187 *both* David Purdie
p.190 *top* Roger Ressmeyer, Starlight/Science Photo Library; *middle* Jodrell Bank/Science Photo Library; *bottom* Dr K. Milne/David Parker/Science Photo Library
p.191 *article* New Scientist, London; *photo* Leonard Freed/Magnum
p.194 *both* ZEFA
p.196 *top* Doug Allan/Oxford Scientific Films; *bottom left* George McCarthy/Bruce Coleman Ltd; *bottom right* H.D. Brandl/Frank Lane Picture Agency
p.198 *both*, p.199 Ron Boardman/Life Science Images
p.201 *article* Philip Allan Publishers (from *Catalyst*, September 1990); *photo* Associated Press Photo

The publishers are grateful to the following examining groups for allowing reproduction of examination questions:
London and East Anglian Group (LEAG)
Midland Examining Group (MEG)
Northern Examining Association (NEA)

1

Our world

The living world

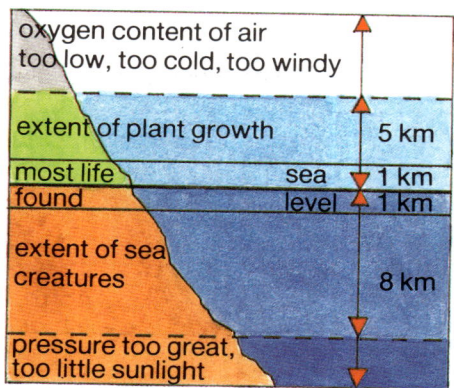

The Earth's biosphere

Where there is life . . .

The part of our planet which contains living things is called the **biosphere**. It extends a few kilometres above and below sea level but only a few metres into the soil. It surrounds the Earth like a blanket, regulating the sun's rays and the movement of gases.

The place where a plant or animal lives is called its **habitat**. This contains all its requirements for food and shelter. A pond is the habitat of the great pond snail. A wood is the habitat of the bluebell.

The terms biosphere and habitat are used to describe *places* where living things, or **organisms**, live. The name given to all the organisms which are found in a habitat is a **community**.

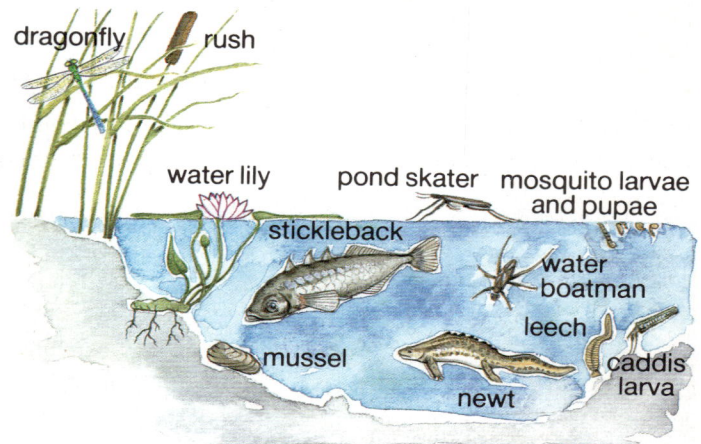

Organisms in a pond community

Organisms within a community are in competition for the food, water and space needed to grow and multiply. This is a delicate balance and it is easily upset by people who, perhaps unintentionally, destroy a habitat. As human communities around the world develop industry and agriculture, greater demands on the Earth's resources mean there are fewer and fewer natural habitats left. However, new habitats are also being created.

A new habitat for wild flowers

In any habitat, such as a pond or a wood, the community is made up of three types of organisms: **producers**, **consumers** and **decomposers**. These organisms make up the **food web** of the community. Each habitat has a unique community of organisms. You would not find a water lily in a wood or a bluebell in a pond, for example, although living things can sometimes adapt to strange environments.

You are likely to find several habitats within walking distance of home and school. These might include a deciduous or coniferous wood, a hedgerow, a stream, a pond, grassland, moorland, marshland, saltmarsh, a shingle or sandy beach.

The British Isles have a wide variety of habitats

A woodland ecosystem

Ecosystems

Living things interact with one another and with the world around them. The term **ecosystem** is used to describe living organisms together with the non-living parts of their habitat and the way they interact. The living, or **biotic**, part includes all the species in an ecosystem. The non-living, **abiotic**, part of an ecosystem includes the climate, the soil, the sun's rays, and the geology of the area. An Amazon rainforest is a large ecosystem; an area of woodland is a small ecosystem.

The study of these relationships is called **ecology**.

Ecologists study ecosystems and the relationships within ecosystems for many reasons, for example:

- to find out why a species is dying out or multiplying
- to find out why an organism is found in some areas but not others
- to investigate the effects of farming, fishing or pollution on organisms in a habitat.

a Biotic parts

b Abiotic parts

Surveying an area

A **quadrat** is a simple and useful way of investigating how many plant species are in an area. It is a wire frame which is divided into smaller squares using string. The quadrat is placed in an area at random. It is usually thrown over the shoulder (after checking that no-one is near!) This means that the thrower does not actually select a place. Then either the number of individual plants in each square is counted (difficult if it is grass) or the percentage area covered by a particular species is estimated.

A **transect** is a line stretched between two pegs. This can be used in two ways. The number of plants directly beneath the line can be counted, or a series of quadrat surveys can be taken along the line.

Beating or **sweeping** is a way to find out how many species of small animals inhabit a tree or a group of bushes. A large cloth is placed on the ground under the tree and the branches above are beaten with a stick. A large sweep-net waved over bushes or flowers will catch insects such as hoverflies and butterflies.

Abiotic factors which affect the organisms living in an area include:
- soil—type, temperature, water, humus and mineral content
- air—temperature, humidity (proportion of water in the air)
- light—intensity
- wind—direction, strength, frequency of gusts
- pollution—litter, chemicals, industrial waste.

Collect
- quadrats
- transect apparatus
- pH meter
- thermometer
- any other apparatus you need

IT Oxygen, light and pH can be measured using a monitoring kit connected to a microcomputer. The results can be stored and processed later.

Investigate a local habitat
Choose one of the following:
a plants along a slope
b an area in/out of shade
c a line 10 metres from a tree.

Design a project which investigates the type of plant species in your chosen area. You might consider the following.

1 Leaf type (e.g. long and thin, short and rounded, smooth-edged, not smooth-edged); leaf orientation (i.e. towards or away from the light).
2 Biotic factors such as nearby trees, other animals, human activity.
3 Abiotic factors, for example drainage, protection from/exposure to wind, amount of sunlight per day, etc.

You will need to write down:

- which variables are to be studied
- how these variables are to be recorded
- how any comparison can be made valid.

Carry out your investigation.

Q1 Present a report of your investigation, which should include:
a a title
b a description of how you carried out the investigation
c an appropriate way of displaying your results (perhaps draw the main plant characteristics that you used to distinguish the plant species and list the number of plants with a certain characteristic).
d a conclusion.

 *Take notes*

Discuss with a partner and write down:

- a suitable heading for your notes
- a sentence explaining each of the following words: biosphere, habitat, organisms, ecosystem, biotic, abiotic, ecology
- three methods of surveying an area
- how humans or pollution may have a bad effect on a community.

Life in a habitat

Oakwoods are very versatile—they can grow on both sandy and clay soils. They also tolerate very cold and wet winters. In summer the trees offer shade for wildlife and in winter protection from frost. This means that temperatures are less extreme in the wood than in the surrounding countryside.

Plants and shrubs that grow in the wood tend to be shade loving or early flowering. They flower early in the year because there are fewer leaves to block the sunshine. There are known to be nearly 300 insect species that live and feed on the roots, bark and leaves of oak trees. (There would be ten to twenty times fewer in a pinewood.) In turn the insects attract many birds.

An oak tree is an example of an ecosystem. It provides shelter and food for hundreds of different organisms, at every level:

- crown—birds and squirrels build nests
- branches and barks—ivy, mistletoe, lichens, mosses, algae, fungi grow
- leaves—eaten by wasps, moths, beetles, weevils
- acorns—eaten by birds, insects, mammals
- roots—inhabited by weevils, earthworms

An oak tree and its community

1 Collect resource sheets and use the information to make up five food chains for the oak tree. Label the primary and secondary consumers, producers and decomposers.
2 Combine the food chains into a food web.

Q1 Why do you think that oak trees are not overwhelmed by the great numbers of organisms they house?

Q2 The oak tree is so adaptable that there are over 450 types of oak tree in the world, but only two are native to the British Isles. Find out (from a field guide or encyclopedia) which the native species are and what the main differences are between them.

[IT] You could use a 'tree-structure' or key on the computer to identify the trees or use 'Branch' to construct your own key.

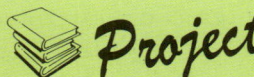

Project

Go for a walk in a local wood or park with a group of friends. Make a sketch of a part of it showing the different layers. Use a field guide to identify some of the plants in each layer. Note down any animals that you observe and where they are.

Populations

Human influences on plant and animal populations

Habitats are not fixed—they change over a period of time. Some change very quickly while others change over a long period of time. These changes may be caused by changes in the weather or by the plants and animals which live in the area. However, the most important cause of change is human influence. Even a minor change can affect several food webs.

Many farmers consider foxes a nuisance because they sometimes kill sheep and chickens. Foxes also eat grey squirrels, so if farmers kill the foxes the number of squirrels increases. Squirrels damage trees, so killing the foxes has an adverse effect on trees which are an important part of the ecosystem.

If the population of one of the organisms in a food web is altered then all the others are affected. In 1954 an epidemic of myxomatosis killed nearly all the rabbits in Britain. This upset the balance of nature in the countryside. Animals which eat rabbits were affected: buzzards became fewer but foxes ate more poultry and voles. Large areas of grassland previously cropped by rabbits became overgrown with shrubs. Since then the rabbit population has gradually recovered.

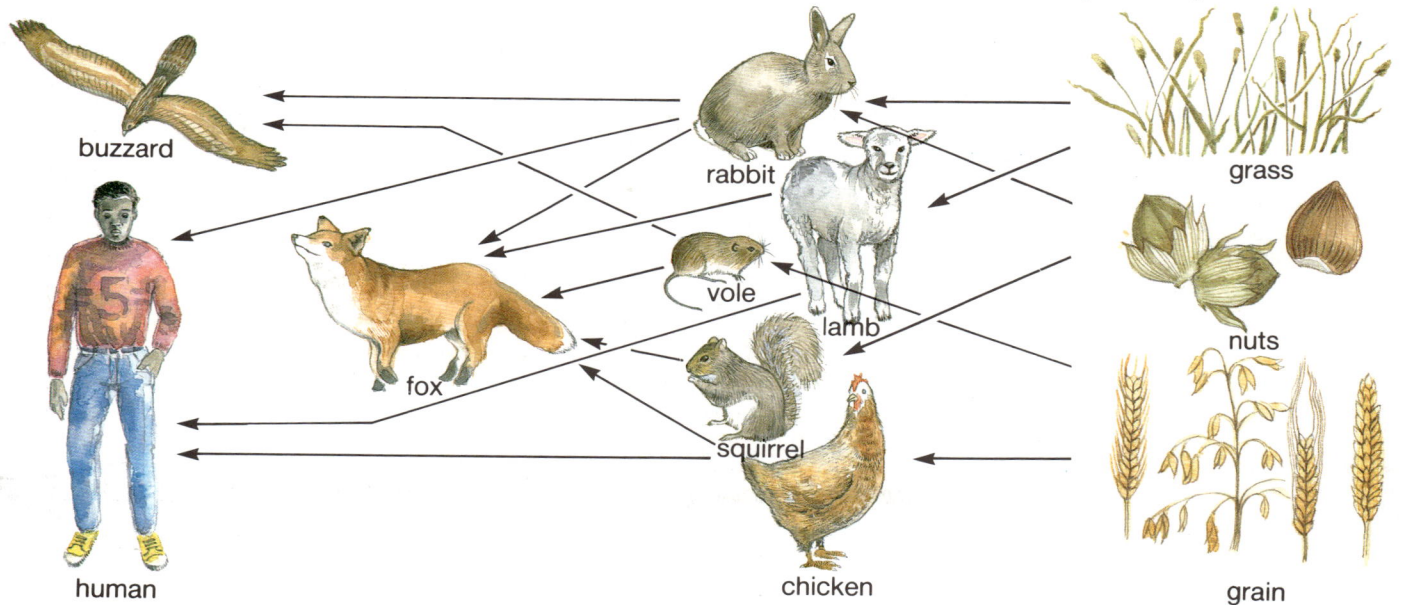

buzzard
rabbit
grass
human
fox
vole
lamb
nuts
squirrel
chicken
grain

The Grand Canyon (USA) is a national game reserve. In 1910 it was thought that the deer population needed to be protected. This was done by shooting the animals which preyed upon them. These included coyotes, cougars, wolves and bobcats. Fourteen years later the deer numbers had increased from 4000 to 100 000. The grassland was overgrazed and young shrubs were destroyed.

Q1 Explain why killing foxes may have an effect on trees.

Q2 Which animals (apart from humans) are most commonly seen where you live?

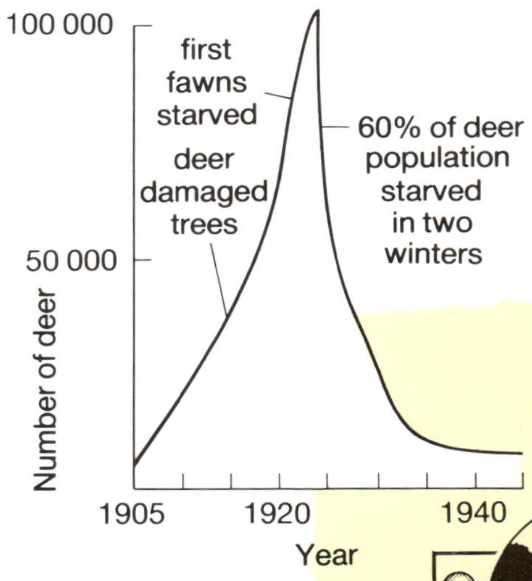

Number of deer (y-axis)

100 000

first fawns starved

deer damaged trees

60% of deer population starved in two winters

50 000

1905   1920   1940

Year

Q3 Draw a food chain for the rabbits which were in danger from myxomatosis.

Q4 Use the graph of the Grand Canyon deer population on the left to explain what effect human interference had on the food web and the deer numbers.

Human population

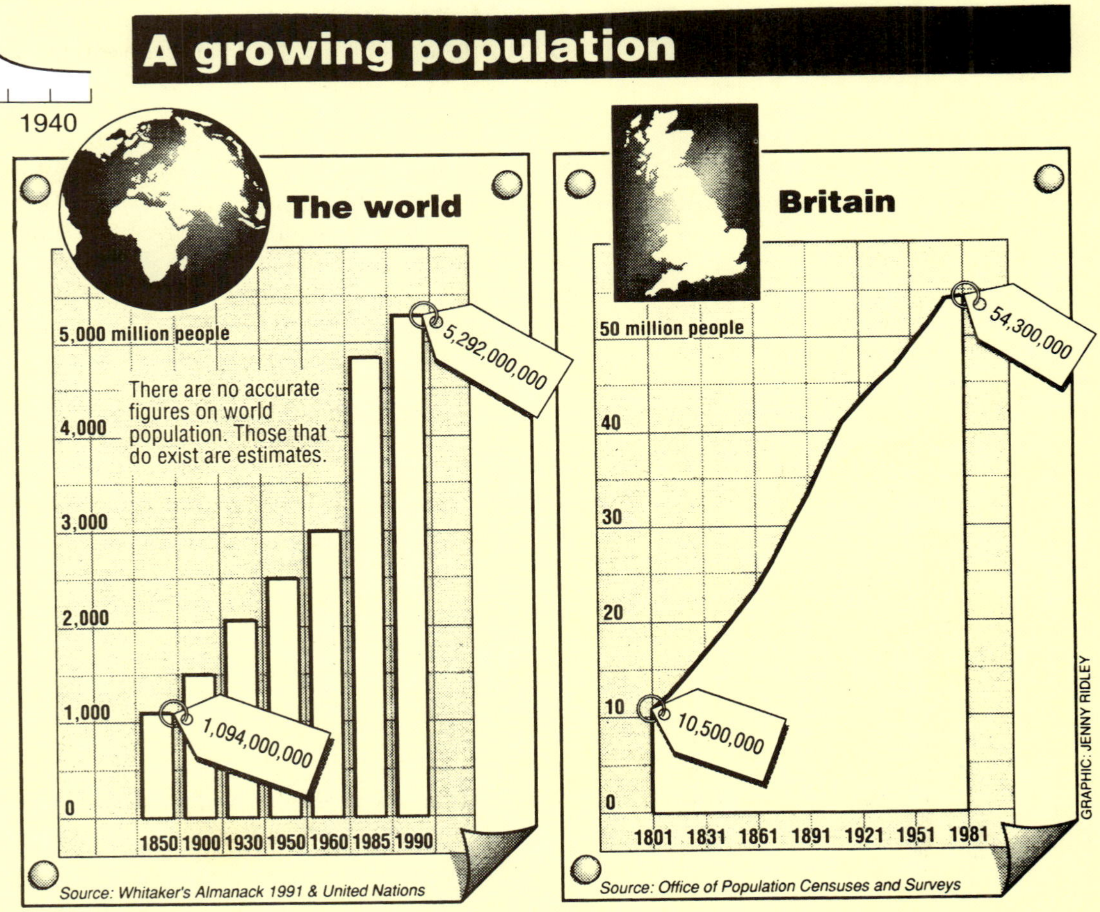

A growing population

The world

5,000 million people

There are no accurate figures on world population. Those that do exist are estimates.

4,000

3,000

2,000

1,000

0

5,292,000,000

1,094,000,000

1850 1900 1930 1950 1960 1985 1990

Source: Whitaker's Almanack 1991 & United Nations

Britain

50 million people

40

30

20

10

0

54,300,000

10,500,000

1801 1831 1861 1891 1921 1951 1981

Source: Office of Population Censuses and Surveys

GRAPHIC: JENNY RIDLEY

It is estimated that over the last 100 years the population of the world has more than trebled. During this time the population of Britain has doubled. This means that more land is needed for living, working and travelling. More people choose to live outside cities, which means the countryside is gradually getting built up. Covering the landscape with concrete means that there are fewer habitats for many animals and plants. Of course some animals and plants have adapted to city living. For example, many foxes now live in towns and the number of magpies to be found in towns and cities has increased enormously over the last twenty years.

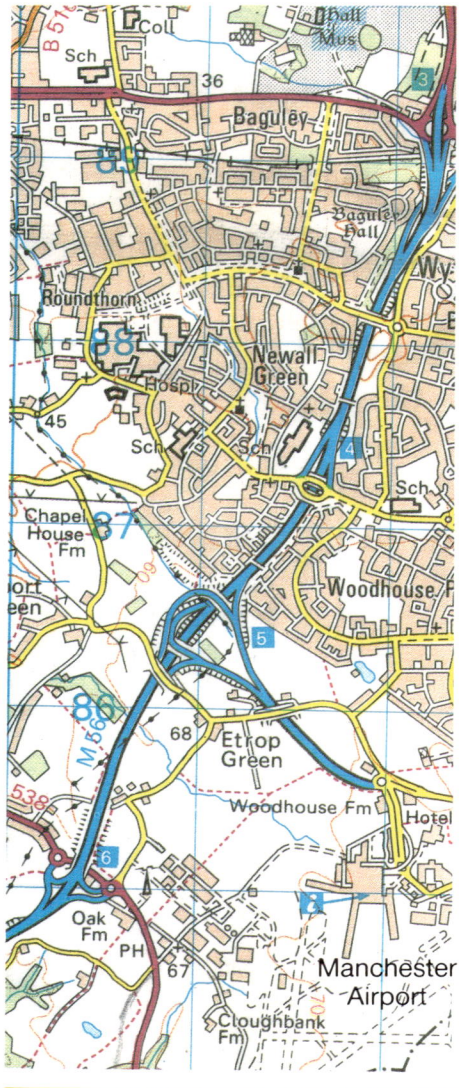

An area near Manchester in 1924

The same area today

Q5 Study the two maps and discuss with your partner:

- Which habitats have been completely destroyed?
- Which new habitats have been created?

Draw a table to show the ways in which the landscape has changed since 1924.

Design and carry out an investigation into the effects of people walking in an area

What effect does trampling the land have on the plants which grow there or on the animals which are found in the soil?

You must consider and decide with a partner:

1 Which variables are most important?
2 How are these variables to be measured and recorded?
3 How can any comparison be made fair?

Q6 Present a report about your investigation.

- Describe how you carried out the investigation.
- State which variables were controlled.
- Include diagrams and tables of results where appropriate.
- Give your conclusion.
- Mention your opinions about the effect of human influence on the area.

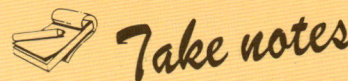 Take notes

Discuss with your partner and write down:

- a suitable heading for your notes
- a list of factors which affect habitats
- an example of human interference in an ecosystem.

Population growth

Animal and plant populations are affected by changes in both biotic and abiotic factors.

Collect
- seeds
- containers
- compost

Compare the germination and growth of seeds which are sown at different densities

Remember to select all your seeds from the same packet.
Remember to keep all abiotic variables the same.

Q1 Record your results in a table, and draw a graph if possible.

Q2 What can you conclude from your investigation? Give reasons for your answer.

 Project

Select a small area near to school or home.
Visit the local library and try to find out what the area was like 100 years ago.

[IT] Use TTNS or Campus 2000 to find this information and compare your area with other areas.

An underwater existence

The oceans cover nearly three quarters of the Earth's surface. The three major oceans are the Pacific (which covers one third of the Earth); the Atlantic, and the Indian Ocean. The deepest part of any ocean is the Marianas Trench in the Pacific, just east of the Philippines. It is 11 000 metres deep. This is deeper than the height of Mount Everest.

Water provides a protective environment for many plants and animals, as it prevents the **desiccation** (drying out) of external body tissues, provides buoyancy, transports food and dissolved gases essential for respiration to the organism and waste materials and gases away. It also makes fertilisation easier than on land by providing a transport medium for the **gametes** (male sperm and female eggs), prevents desiccation, helps the young to disperse and protects them from the harmful ultraviolet rays of the sun.

One of the most important properties of water is that it is a very powerful **solvent** (dissolving agent). The water provides all sorts of dissolved minerals such as salts and nutrients as well as gases that are essential to sustain life. This mixture of water and dissolved substances is called a **solution**. The minerals and salts are eroded from rocks and soils by the mechanical action of the sea, or are released from dead organisms by the action of decomposers.

The water which evaporates from the sea to form rain is pure water. The dissolved salts in the sea are left behind during the process of evaporation. High **salinity** (high density of dissolved salts) is found where rainfall is low but evaporation is high. This is why the Dead Sea is very salty.

Marine or saltwater ecosystems include mangrove swamps, coral reefs and estuaries as well as oceans

Ocean ecosystems

These can be divided into two main types: **coastal ecosystems** and **ocean deeps**. A coastal ecosystem can be up to 200 m deep and extend from the water's edge at low water to the edge of the **continental shelf**. (This is the region of the sea bed closest to land. It is about 100 km wide.) This region contains 98% of ocean life forms even though it occupies less than 10% of the ocean area.

Most commercial fishing and off-shore oil platforms are located in the coastal ecosystems and these are becoming over-fished and increasingly polluted. The deep sea areas are relatively unaffected at the present time.

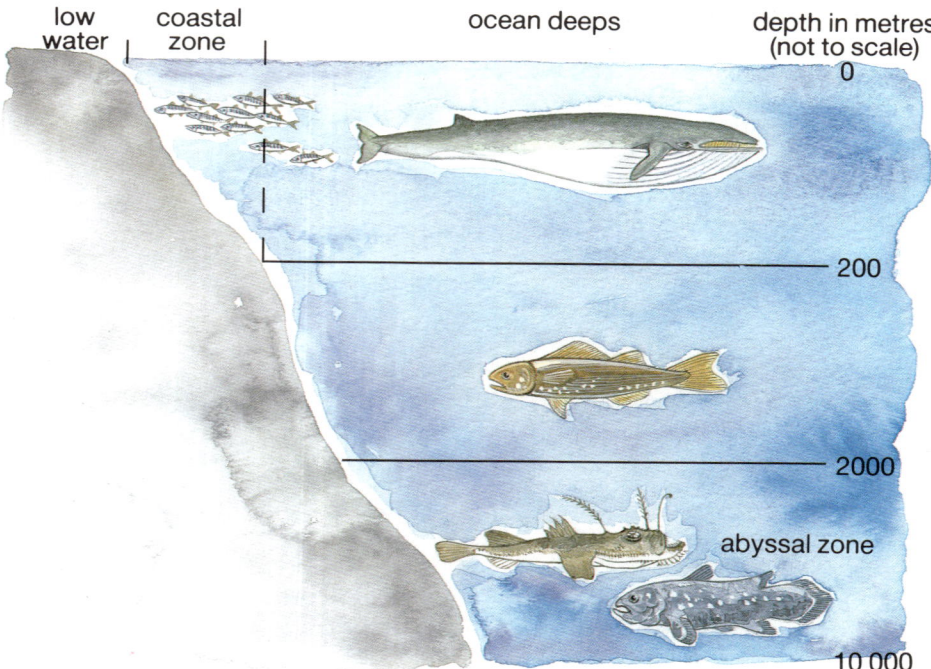

low water — coastal zone — ocean deeps — depth in metres (not to scale)

0

200

2000

abyssal zone

10 000

Freshwater ecosystems

Freshwater ecosystems include marshes and wetlands, ponds, lakes and reservoirs, rivers and streams. They usually have very low salinity so support different species of plants and animals.

These ecosystems vary a great deal because of differences in geology, pollution and use of the land. In most freshwater ecosystems there is a build-up of nutrients and of plant material. This is called **eutrophication**. This process is speeded up when sewage or agricultural waste such as fertiliser is added to the water. The oxygen content of the water is reduced (see diagram) and this has a harmful effect on the food web.

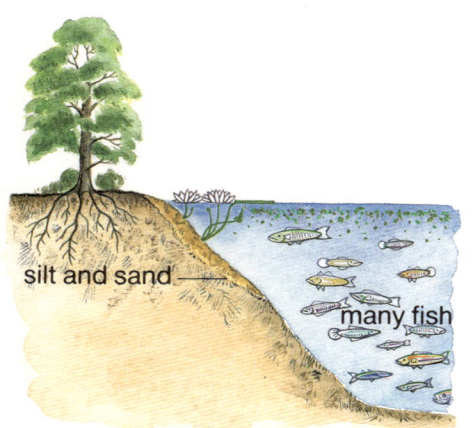

Nutrient-rich lake (eutrophic)

heather moorland and/or grassland

Nutrient-poor lake (oligotrophic)

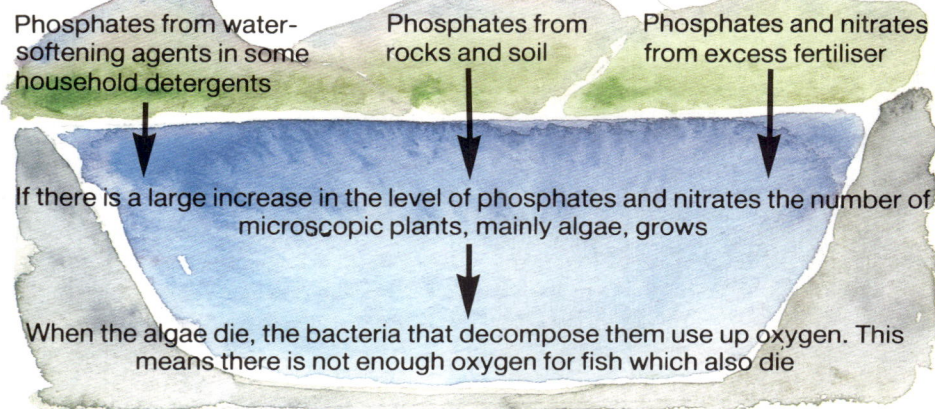

Phosphates from water-softening agents in some household detergents

Phosphates from rocks and soil

Phosphates and nitrates from excess fertiliser

If there is a large increase in the level of phosphates and nitrates the number of microscopic plants, mainly algae, grows

When the algae die, the bacteria that decompose them use up oxygen. This means there is not enough oxygen for fish which also die

Eutrophication cycle in a nutrient-rich freshwater ecosystem

Sometimes the water comes from an area of acid rocks or peat and this means the water has very little nutrient content.

Collect
- 2 samples of water
- 0.1% methylene blue solution
- pipette

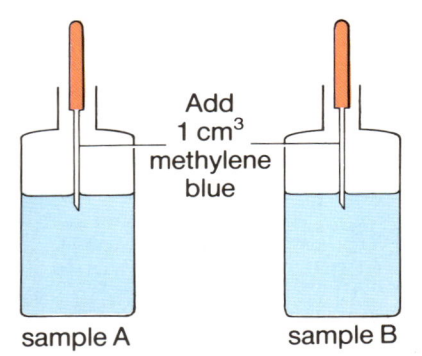

Add 1 cm³ methylene blue

sample A   sample B

Compare the oxygen content of different water samples
- Replace stoppers immediately after adding the indicator solution.
- Put the bottles in a warm, dark place.
- Note the time it takes for the blue colour to disappear. If there is no change by the end of the lesson, check daily.

The longer it takes for the blue colour to disappear, the higher the dissolved oxygen content.

Q1 Draw up a table to compare the two types of freshwater ecosystem shown above.

Q2 Which of your samples would support the most life?

Q3 What is eutrophication?

Q4 What causes eutrophication to speed up?

Q5 How could eutrophication be reduced?

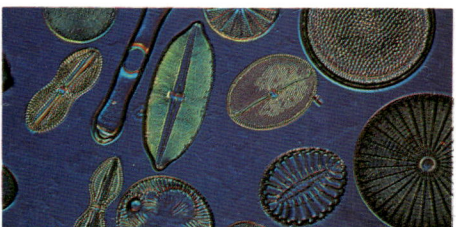

Phytoplankton ×115

Zooplankton ×350

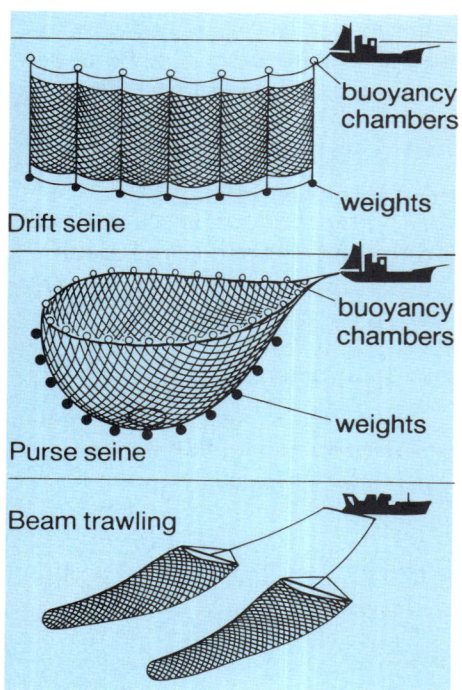

North Sea fishing

Over-fishing is thought to be an even greater threat to the North Sea than pollution. Commercial fishing is affecting the balance of life and some species are in danger of dying out because the food web is being disturbed.

The primary producer in the sea is plant plankton, **phytoplankton**, and the primary consumers are planktonic animals, **zooplankton**. Plankton is the most important food for many sea animals such as herring and sprats. These in turn are food for seals, dolphins, sharks and rays. Starfish and crabs feed on the dead and dying fish thrown back by fishermen as well as on mussels and other dead animals.

Commercial fishing techniques

The two main techniques are **drift** and **trawl** fishing.

Drift techniques, such as **drift seine** and **purse seine** shown left, are used to catch surface-living fish such as herring, mackerel, anchovies and sprats. The nets are dragged slowly through the water or pulled tight so that all the fish, young and old, are trapped.

Trawling is used to catch fish that live near the sea-bed, such as plaice, cod and haddock.

Modern fishing vessels have electronic and sonar fish-finding equipment. Shoals of fish can be detected, whatever their depth, and the size and type of fish can be estimated.

Fishing under threat

These efficient methods of fishing are not good for the fish species. The more fish caught, especially young fish, the fewer there will be in the future. Cod and haddock, for example, should live for at least ten years. However, fewer than one third survive for more than twelve months. This means many do not reach breeding age and so dwindling stocks are not replenished.

This is not good for the future of the fishing industry, either. However, reducing the numbers of fish caught (the 'quota') would mean that many fishermen would go out of business. This could affect entire communities, such as the port of Peterhead in north-eastern Scotland, which rely on fishing to survive.

Dolphin disaster

There is a danger that the dolphin, which is a beautiful, friendly and intelligent mammal, will die out because its food supply is being reduced by over-fishing.

In addition, more than half a million dolphins have been killed each year worldwide, due to the use of enormous drift nets in the south and eastern Pacific which were intended to catch large fish such as tuna. The use of these drift nets was finally banned in 1991.

Q1 Draw a food web of the organisms illustrated here.

fulmar
haddock
cod
herring
dolphin
starfish
sprat
hermit crab
herring gull
plankton
seal

Q2 Summarise the action taken so far to solve the problems of over-fishing and suggest some of your own.

Q3 Explain why starfish and crabs have increased in numbers in spite of the fishing industry.

Q4 Advances in fish-catching technology have reduced the number of haddock in the seas around Britain from 122 billion to 7 billion in less than 20 years. Explain how this has happened.

IT Use TTNs or Campus 2000 to find the most up-to-date information on EC fishing regulations.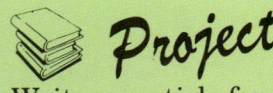

Project

Write an article for your local newspaper about the effect of commercial fishing on dolphins. Use diagrams or pictures to illustrate your points.

Action taken so far to stop over-fishing

1976 The International Council for the Exploration of the Sea banned herring fishing for several years.
1977 The European Community banned fishing within 320 km around their coasts.
1982 The United Nations set up a Convention on the Law of the Sea, which 160 countries have signed. This includes an Exclusive Economic Zone of 364 km off the coast of each country, to protect both fishing and mineral rights.
1983 The European Commission set fish quotas for European Community countries, to be revised in 1992.
1986 The European Commission recommended that countries cut their fishing fleets by 3%. This was largely ignored.
1990 The quotas for British fishermen were reduced by 20 000 tonnes for haddock and 15 000 tonnes for herring. The European Commission demanded that the traditional diamond-shaped trawl mesh be replaced by a square shape which does not close up under pressure. This would allow the younger fish to escape and breed. Fishermen argue that this is difficult to do with mixed fishing and means that they catch fewer whiting which are not under threat.
1991 The use of ocean drift nets was banned.

To rake or not to rake?

Some gardeners prefer to rake the grass after cutting, others say that leaving grass cuttings where they are is beneficial to the lawn.

Your problem is to investigate which method is likely to be best for the lawn

Work in a group.

You will need to consider:

- which variables you should keep constant
- whether to use two similar plots or to conduct an experiment in the laboratory
- how you will measure or determine the benefit to the grass.

Discuss and decide on the best method. Write down an outline plan and equipment list.

Carry out the experiment. Record your observations over a period of a few weeks.

Q1 Discuss your results and produce a report, including your conclusions.

[IT] Use a database to process and analyse your results.

Q2 Send your report to a gardening magazine such as Gardening *Which?*

Remember to include diagrams and photographs to support your conclusions.

M6 outcry

The Department of Transport plan to replace the 7-acre motorway services at Knutsford in Cheshire with a 50-acre service area in Green Belt land. There is a considerable public outcry at this 'unnecessary and unwarranted intrusion into Green Belt land'.

Work in a group.

Collect resource sheets of newspaper cuttings and maps. Study these and the pictures and article here.

- Make a list of the people who are objecting to the proposals.
- Discuss the points they make.
- Discuss why Green Belt areas are important.

M6 services – deadline approaches

TIME is running out for people wishing to protest about plans for a controversial new motorway service station to be built on land near Cann Lane, Aston-by-Budworth.

The application for outline planning permission will be discussed by Macclesfield Borough's main planning committee on Monday, April 8, and Clr Moira Taylor is urging anybody opposing the plan to write to the council as soon as possible.

Green Belt

"This is 50 acres of prime Cheshire green belt land," said Clr Taylor. "It is a much-loved and well-used beauty spot, attracting lots of visitors.

"The development will be much bigger and far more obtrusive than the current one at Knutsford, incorporating 1,100 cars, 300 lorries, 50 caravans, with overnight hotel facilities, picnic areas and petrol stations."

Aston-by-Budworth Parish Council is also protesting strongly against the plans. They say it would make more sense to build a new service area somewhere halfway between Sandbach and Charnock Richard, possibly in the Woolston area, making the services in this area about 20 miles apart, and saving the green belt.

And Clr Bob Ingham has also strongly objected to the proposals. He said:

Disrupt

"In my opinion, there is no planning logic in the location of the new service centre, because the suggested site is in the middle of a quiet rural community and building operations will totally disrupt the whole area.

"If the location was moved to Junction 20, an area already "spaghetti junctioned", it would be in a position to serve the M56, with no services, the M6 and any future Thelwall Viaduct expansion."

Knutsford Guardian, 20.3.91

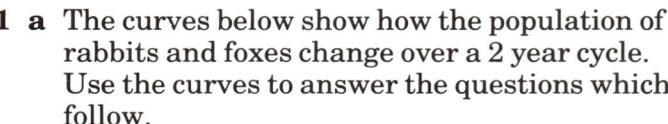

1 a The curves below show how the population of rabbits and foxes change over a 2 year cycle. Use the curves to answer the questions which follow.

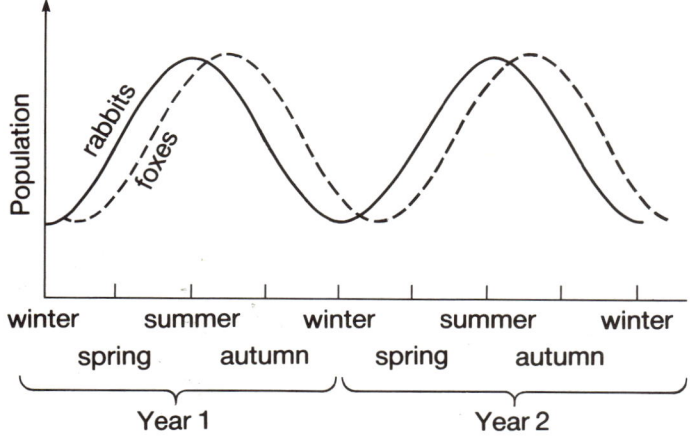

i At what time of year is the rabbit population lowest? *winter* [1]

ii At what time of year does the rabbit population rise most rapidly? *summer* [1]

iii Why do you think that the fox population falls each year? *because rabbit pop goes no food* [1]

b Use the information given in the simple food web below to answer the questions which follow.

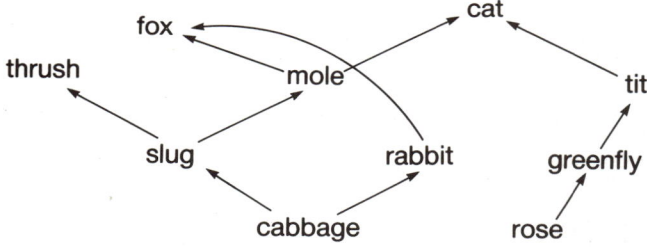

i Name: a major predator, a herbivore, a carnivore and a producer. *fox, rabbit cat cabbage* [4]

ii Complete the following food chain.

cabbage → rabbit → ? *fox* [1]

iii What would be the effects of removing the foxes? *to many moles + rabbits Overcrowding + death Cats have more food* [3]

MEG 1988 Specimen Paper Science Syllabus B

2 The following table gives the total dry mass of plant material produced per year in different types of habitat in Britain. In each case, it is assumed that none of the plants has been consumed by animals.

Habitat	Dry mass of plant material (per m² per year)
Grassland	600
Cereal crops	2200
Deciduous forest	1200
Coniferous forest	2800
Freshwater pond	600
Salt marsh	3000

a Represent this information by means of a suitable chart. [4]

b i Which one of the habitats given in the table would support the largest population of animals? *salt marsh* [1]

ii Explain why you chose this habitat. [2] *more food more environment*

NEA Specimen Paper Science Syllabus B

3 The following account is taken from the diary of a biologist on an Antarctic survey ship in 1885.

"The waters around this frozen land contain vast numbers of tiny shrimps known as krill. These feed on the many small plants which occur in the surface waters.

The krill are the main food for squid and whales which arrive in the summer to feed.

The albatross, a large seabird, is also common and feeds on the squid and krill.

The whales are easy prey for whaling ships and the numbers of whales are falling rapidly."

a From the information above, draw a food web for the Antarctic in 1885. [5]

b The diary continues—*"In the Antarctic winter much of the ocean freezes over and light is prevented from reaching the small plants."*

i How is this likely to affect the size of the krill population? [1]

ii Give a reason for your answer. [1]

c In recent years ships have begun to harvest the krill for human food.

i How is this likely to affect the size of the whale population? [1]

ii Give a reason for your answer. [1]

d Suggest **one** way of preventing the whales from becoming extinct. [1]

NEA 1990 Biology (Human and Social)

2
Light

Light reflections

Mirror, mirror

We know light travels in straight lines but what happens when it bounces off things?

When investigating how light rays behave, different names are given to the rays which come from the light source—**incident rays**—and the ones which are reflected—**reflected rays**.

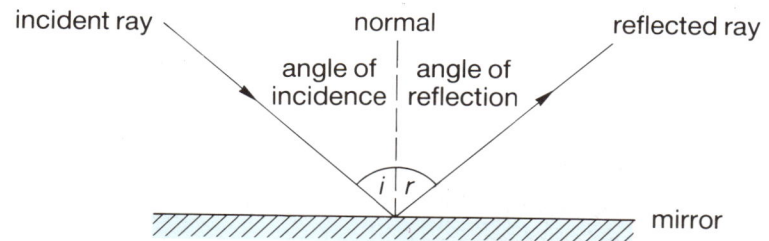

A line can be drawn at right angles to the mirror surface where a ray hits it and this is called the **normal**. The **angle of incidence** is measured between the normal and the incident ray. The **angle of reflection** is measured between the normal and the reflected ray.

Collect

- ray box
- plane mirror and stand
- white paper
- ruler
- protractor

1 Investigate what happens when light is reflected

a Work with a partner. Plan your investigation to show how the position of the reflected ray changes if you move the position of the incident ray.

- What do you expect to happen?
- Write down your ideas in a way that can be tested.
- What angles will you measure to see if your ideas are right?

b Carry out your investigation. You must take the measurements carefully. You can check that they are accurate by repeating each measurement.

c Present your results in a table.

2 How far behind the mirror is the reflection?
You can collect a resource sheet to help you with this investigation.

Curved mirrors

Light can be reflected from curved surfaces as well as plane surfaces, but the effect is quite different.

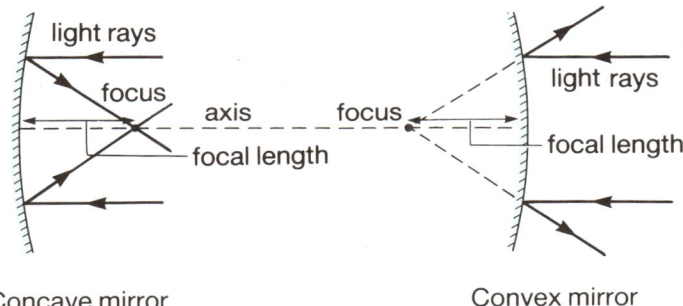

Concave mirror   Convex mirror

Where the light rays cross or seem to have come from is called the **focus** of the mirror. The distance from the mirror to the focus is the **focal length** of the mirror.

Collect

- ray box
- curved mirrors
- white paper
- ruler
- resource sheets if you need help

Investigate what happens when light is reflected from a curved mirror

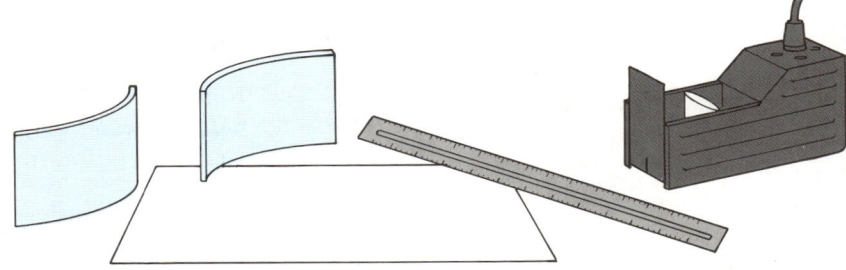

1 a Find the focus of the concave mirror by drawing the reflected rays of several parallel incident rays.

b What is the focal length?

2 Repeat the investigation with the convex mirror.

Treasure hunt

By following the beam of light
you'll soon find my treasure site.
Use the laws of reflection.
Follow every change of direction.
Bounce the ray as it does go
from mirror to mirror, to and fro.
Fifty centimetres you must measure
and there you'll find my buried treasure.

Captain Luminaire

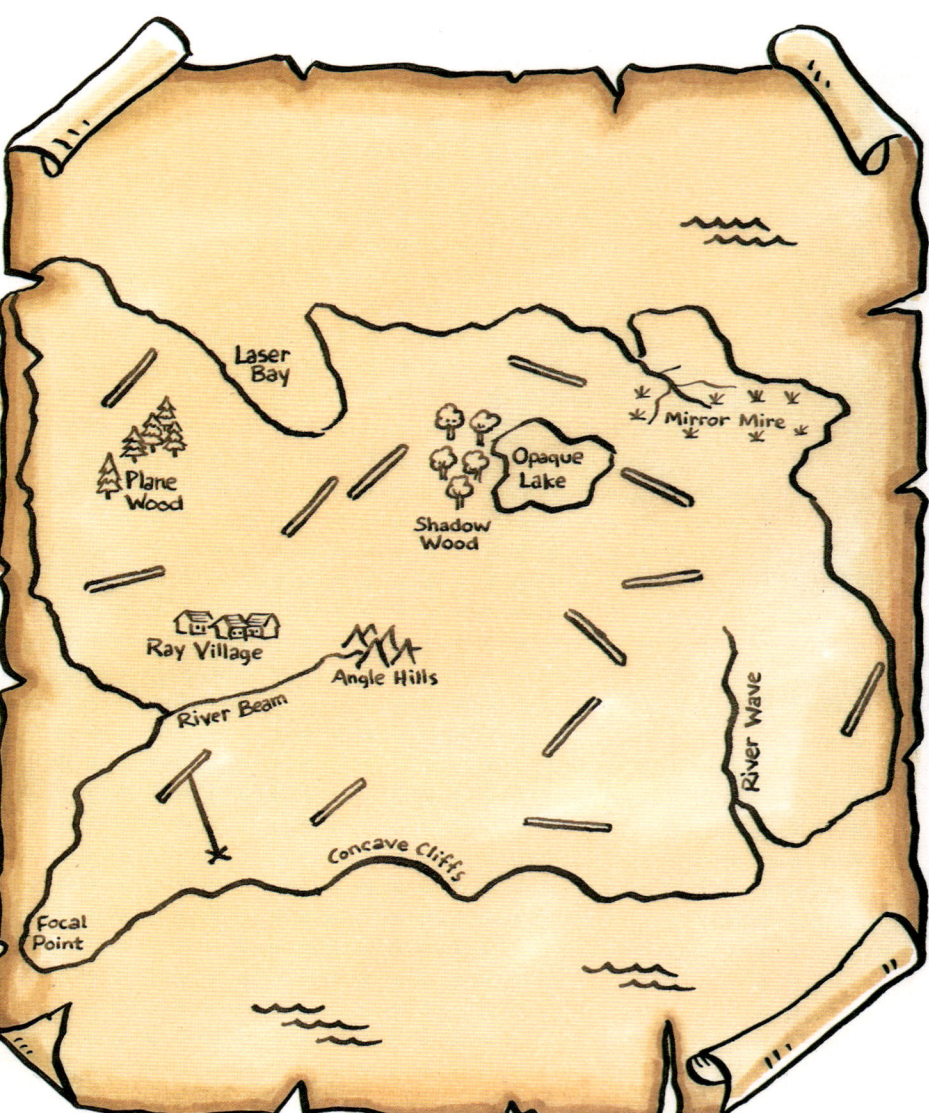

Collect a full-size copy of the map and go treasure-hunting!

Uses of mirrors

There are many uses for plane mirrors. We use them in our homes. They are also used in periscopes, in kaleidoscopes, for signalling, for looking around corners, in meters, in microscopes, and as reflecting number plates.

Curved mirrors also have many uses, as shown in the photographs.

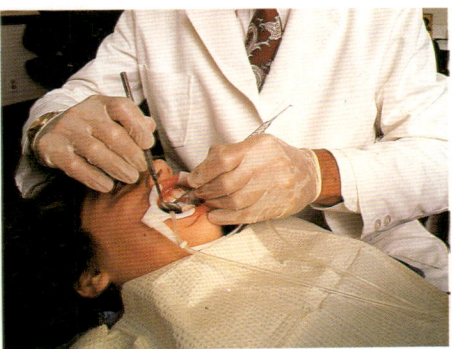

Convex mirrors give a wide-angle view, so are used on dangerous bends of roads . . .

Concave mirrors enlarge a near object, so are used by dentists . . .

. . . for special effects, and for security in shops and on buses

. . . for applying makeup and for shaving

A concave mirror is used in a reflecting telescope to capture the light from a distant object

A concave mirror is used in a microscope to reflect light onto the object

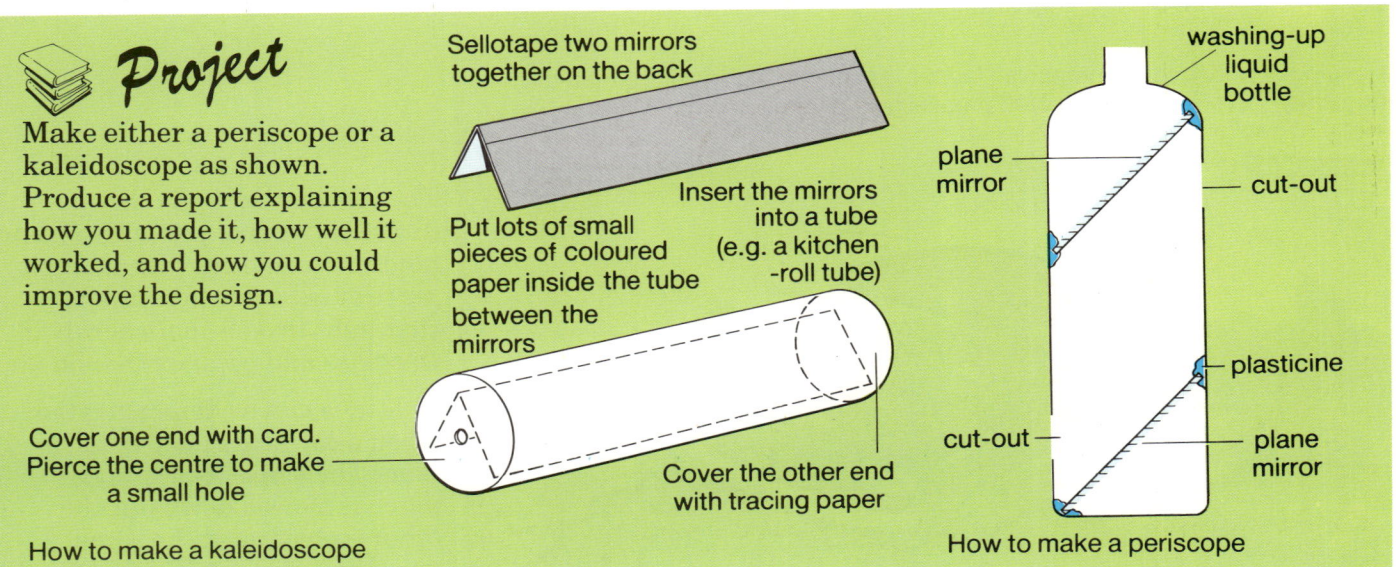

Project

Make either a periscope or a kaleidoscope as shown. Produce a report explaining how you made it, how well it worked, and how you could improve the design.

Sellotape two mirrors together on the back

Put lots of small pieces of coloured paper inside the tube between the mirrors

Insert the mirrors into a tube (e.g. a kitchen -roll tube)

Cover one end with card. Pierce the centre to make a small hole

Cover the other end with tracing paper

How to make a kaleidoscope

washing-up liquid bottle

plane mirror

cut-out

plasticine

cut-out

plane mirror

How to make a periscope

A change of direction

Why change?

When a ray of light passes into a transparent object such as a prism or a lens, or even water, it usually changes direction. This is called **refraction**.

The amount of 'bending' depends on the angle at which the light ray hits the surface of the object. Lenses make use of this to bring rays of light to a **focus**.

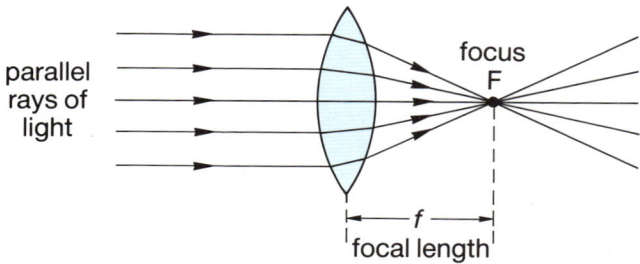

Converging or convex (or 'positive') lenses are fatter at centre than the edge. The refracted parallel rays converge to a focus

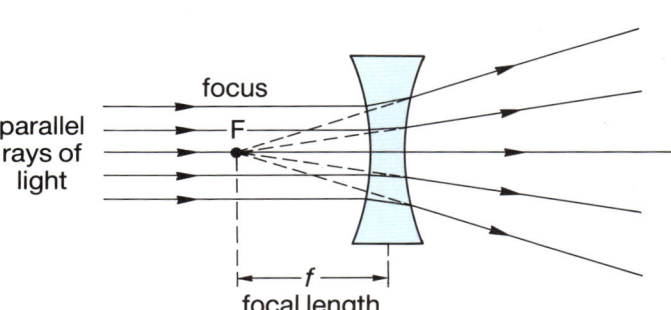

Diverging or concave (or 'negative') lenses are thinner at the centre than the edge (they 'cave' in). The refracted parallel rays diverge and have to be extended backwards to find the focus

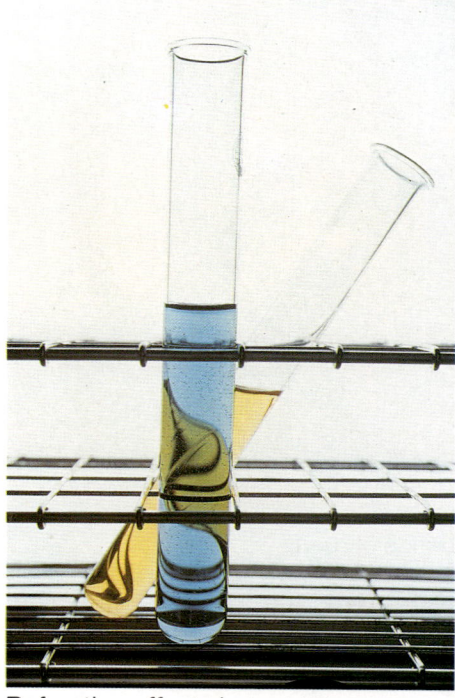

Refraction effects in test tubes

The reason that light changes direction when it passes from one medium to another, for example air to glass or air to perspex, is that light travels more slowly in the glass and perspex than in air.

This idea can be demonstrated when a car hits a deep patch of mud. The wheel that hits the patch is slowed down but the others continue at the same speed so swinging the car round.

Investigate the bending of light

1 Set up the equipment as shown.

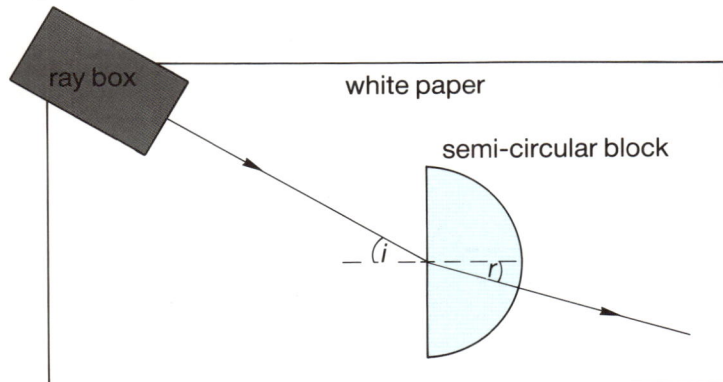

a You will need to compare the angle between the normal and the incident ray (angle *i*) with that between the normal and the refracted ray (angle *r*). Decide how you will measure these angles.

b Move the ray box to a new position and measure the new angles *i* and *r*.

c Repeat several times.

d Record your results in a table.

[IT] A spreadsheet could be used for the results.

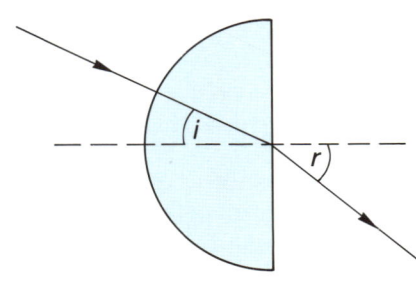

2 Change the position of the semi-circular block so that the light enters the curved surface at right angles (along a radius) and comes out at the centre of the straight edge.

a As before, investigate what happens to the incident and refracted rays and record your results in a table. Use the resource sheet if you need some help.

b Find the angle *i* at which the refracted ray passes *along* the straight edge, neither coming out of the block nor going in.

Q1 In the first investigation what did you notice about the incident and refracted rays?

Q2 In the second investigation what happened when the incident angle was between 40° and 50°? Draw a diagram to show what happened.

Q3 When the refracted ray comes out along the edge of the block the angle of incidence is called the **critical angle**. What is the critical angle for your block?

Q4 What happens to the refracted ray for incident angles greater than the critical angle?

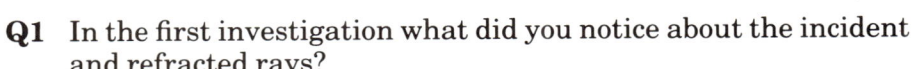

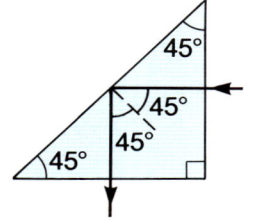

Total internal reflection

We have seen that when the angle of incidence is greater than the critical angle the refracted ray disappears and **total internal reflection** takes place. This only happens when the rays are travelling from a dense medium towards a less dense medium, e.g. glass to air.

A reflector using total internal reflection can sometimes be more useful than a plane mirror. This is because the reflector, such as a prism, reflects more of the incident light than a plane mirror which suffers from the problem of easily damaged silvering. Ordinary mirrors can also give multiple images due to multiple reflection from the glass at the front of the mirror.

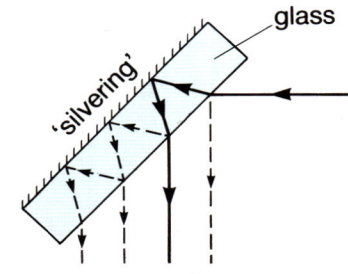

Prisms with angles of 45°, 45°, 90° can be used as reflectors in two ways.

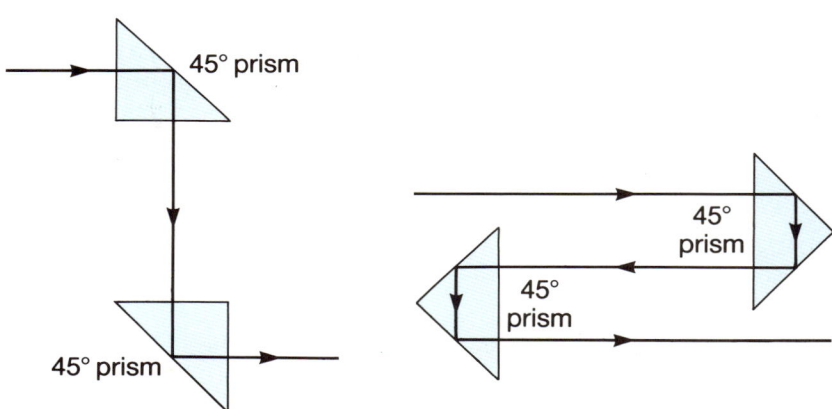

Prisms are used in this way in a periscope

Prisms are used in this way in binoculars and in a bicycle reflector

The prisms in bicycle reflectors turn the light through 180°. Light from car headlights is sent back where it came from so that a car driver can see the cyclist easily. Prisms are also used in this way on warning signs on dangerous sections of road.

Spheres can also be used as total reflectors. The light is reflected in much the same way as in the prisms of bicycle reflectors.

Reflecting spheres are used for cat's eyes in the road. They are designed so that when driven over they squash down, are automatically scraped clean, and then spring back.

Cat's eye

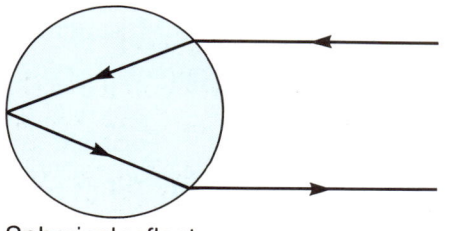

Spherical reflector

Optical fibres

Optical fibres are very thin flexible tubes. They have a narrow core made from pure glass and a cladding (coating), also made from pure glass but with a slightly smaller density. Light rays in the core hit the core–cladding boundary at incident angles greater than the critical angle and are totally internally reflected.

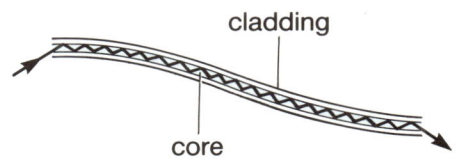

Optical fibres are replacing copper wires in telephone systems. Digitised sound is converted to pulses of light. A pair of fibres is used and one of the many advantages is that each fibre in the pair can carry over 8000 telephone conversations at the same time. Copper wires can only carry 1000 conversations. Optical fibres are also cheaper, lighter, and have no 'cross-talk' between conversations nor interference.

Optical fibres are also used in medicine. They enable doctors to study the inside of the patient's lungs or stomach. Light is sent down them, and the reflected light forms an image of pale and dark spots which can be analysed. If, for example, an ulcer is found, a laser beam can be passed down the fibres to burn and seal it.

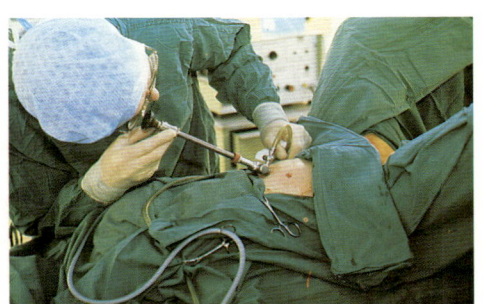

Taking a new direction

1 Investigate how light rays change direction as they pass through the objects provided.

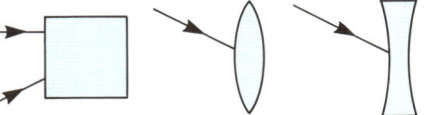
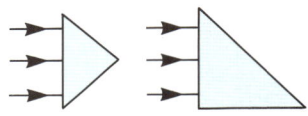

2 Try different combinations of glass blocks. Predict what will happen to the light rays. Test your prediction.

Q1 For each object draw the normal to the surface where the ray strikes it. Draw the normal where the ray emerges from the object.
Where did the ray go inside the object?

Q2 How does the angle of refraction compare to the angle of incidence at the surface where the light enters the object?

Q3 How does the angle of refraction compare to the angle of incidence at the surface where the light emerges from the object?

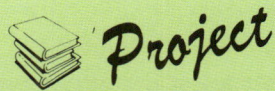

The electromagnetic spectrum

With the speed of light

Light is part of a range of waves, or radiation, called the **electromagnetic spectrum**. Electromagnetic radiation has different properties, depending on the **wavelength** of the wave, but the waves all move forward with the same speed in air.

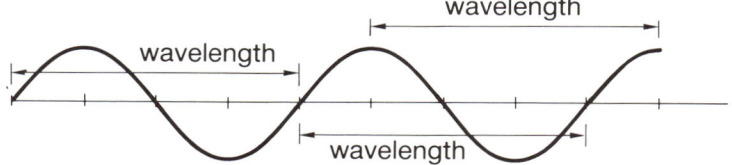

The wavelength is the distance from a point on one wave to the similar point on the next wave

The wave moves forward, and the **frequency** is the number of wavelengths which go past in one second. The shorter the wavelength, the higher the frequency.

Each type of radiation is produced and detected in its own special way. The diagram below shows how the waves are 'spread out' according to wavelength into a spectrum.

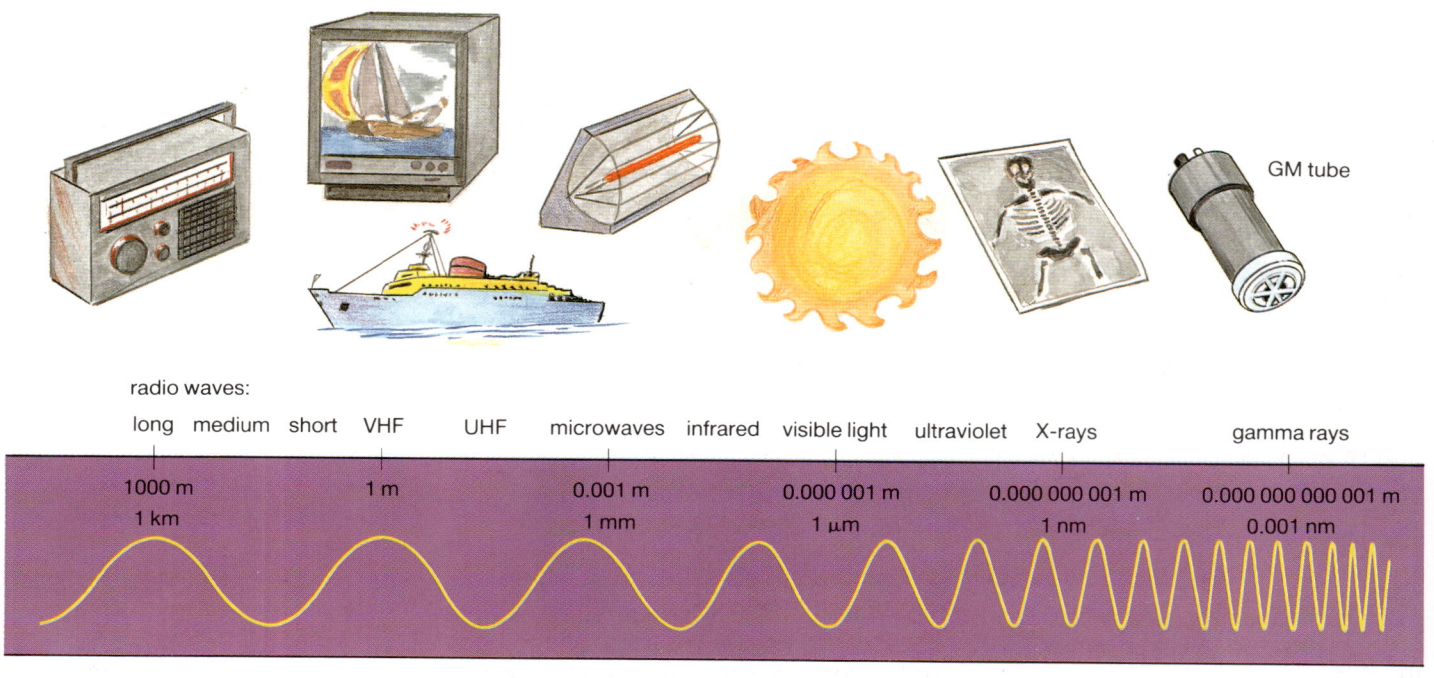

GM tube

radio waves:

long	medium	short	VHF		UHF	microwaves	infrared	visible light	ultraviolet	X-rays	gamma rays
1000 m				1 m		0.001 m		0.000 001 m		0.000 000 001 m	0.000 000 000 001 m
1 km						1 mm		1 µm		1 nm	0.001 nm

Electromagnetic waves have several common properties:

- They can all travel through empty space.
- They all travel through space at the same speed, i.e. the speed of light, which is 300 000 000 m/s or 300 000 km/s.
- They are all ripples of electric and magnetic energy.

Radio waves have the longest wavelength in the electromagnetic spectrum. They are produced from fluctuating electricity in an aerial called a transmitter. The radio waves can be detected by a receiving aerial. The signals received by the aerial are weak and require amplification. They still cannot be seen or heard until the energy they carry is converted by a TV or radio into light and sound. Long, medium and VHF (very high frequency) waves are used to transmit radio, and UHF (ultra high frequency) is used for television.

Microwaves are very short radio waves which are used for **radar** detection of aircraft and ships, and for transmitting television and telephone calls via satellites. They are also used in microwave ovens. When microwaves enter food, some are 'absorbed' by the water molecules in the food and produce heat. This allows the food to be cooked.

Radio transmitting aerial

Infrared radiation from an electric fire

Special photographic film sensitive to IR detects temperature differences and can be used to take pictures in the dark

Infrared (IR) radiation is given off from objects such as fires and radiators. It is absorbed by the human body so we feel it as heat. The radiation is produced by rapidly vibrating molecules.

Most things give off some infrared radiation. The hotter the object, the faster the molecules vibrate and the shorter the wavelength of the infrared radiation becomes. Eventually the object becomes 'red hot'—some of the radiation becomes visible to the eye as red **light**.

reflected

absorbed

pass through — revolving stand

food sample

Microwave oven

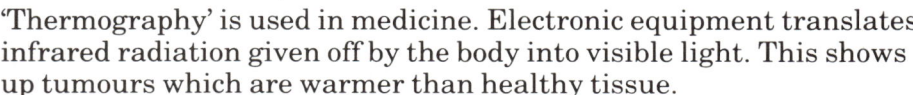

Clothes fluorescing at a disco

'Thermography' is used in medicine. Electronic equipment translates infrared radiation given off by the body into visible light. This shows up tumours which are warmer than healthy tissue.

Ultraviolet (UV) radiation cannot be detected by the eyes, but there is plenty in sunlight and it is this which causes some skins to tan. Excess can be harmful for the skin.

Fluorescence occurs when certain chemicals absorb UV and re-radiate the electromagnetic energy as visible light (they 'glow'). This is used by the manufacturers of some washing powders to make clothes look 'whiter than white'.

X-ray photography has revolutionised medicine. Low-energy X-rays are transmitted by most body tissue but reflected by bone. They affect photographic film, so an image of the bones can be obtained.

A CAT (computer axial tomography) scanner is an improvement over normal X-ray photography. The patient lies with his/her body within a hoop and X-rays are beamed through one side of the body. The beam is picked up by detectors on the other side. The source is moved around the hoop and a picture is built up of a 'slice' of the body. Much greater detail is revealed in this way and differences between types of tissue can be seen, enabling detection of tumours.

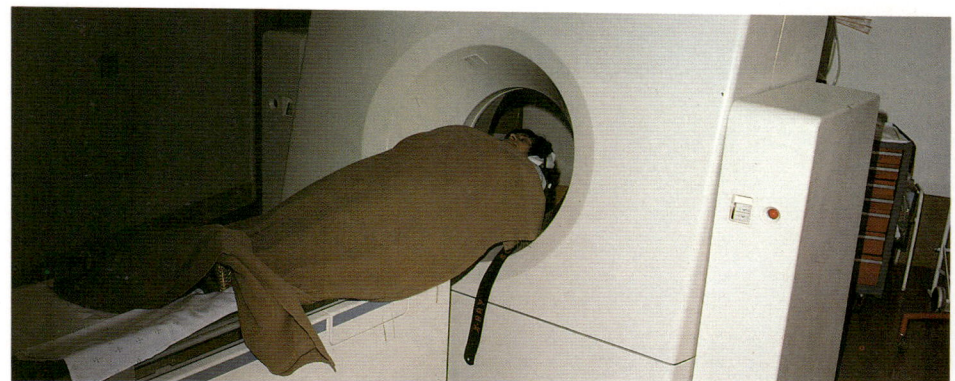

A CAT scanner in use

X-rays can be dangerous, however. High-energy X-rays penetrate the body and can damage body cells.

Gamma rays are given off by radioactive materials and are extremely dangerous because they penetrate deep into the body and damage cells. They are used to destroy cancer cells although it is very difficult to kill the cancer cells without damaging healthy cells nearby.

Hazard warning symbol for radioactive material

Electromagnetic circus
Work in a group. Go round the room, making sure you visit each of the experiments. Follow the instructions and answer the questions on each card.

Q1 Write a report, giving your observations for each type of radiation.

Q2 Name a type of radiation which can:

- cause a suntan
- cause fluorescence
- be used for radar.

Q3 Frequency is measured in Hz (Hertz).
Speed in m/s = frequency in Hz × wavelength in m
and is 300 000 000 m/s for all the electromagnetic spectrum. Use this information to calculate the wavelength of the waves being picked up by this radio.

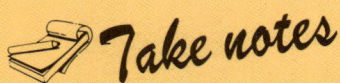

Take notes

Discuss with your partner and write down:

- a suitable heading for your notes
- a table showing *wavelength*, *use* and *name* for the seven types of electromagnetic wave
- what is meant by *wavelength*, *speed* and *frequency* when describing waves.

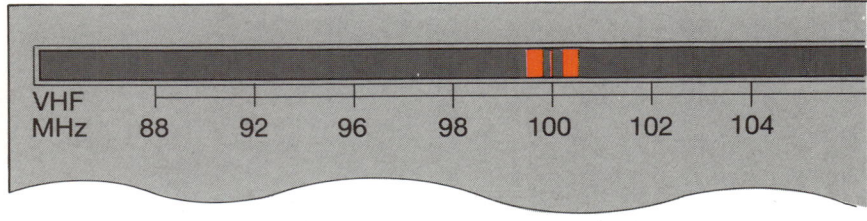

| VHF MHz | 88 | 92 | 96 | 98 | 100 | 102 | 104 |

1 MHz = 1 000 000 Hz

Q4 List some of the ways in which different waves are used for communication.

Q5 Apart from the manufacturers of washing powder, who else might want to use the fluorescent effect that ultraviolet light has on certain chemicals?

Build a radio

Use the kit provided and the instruction card to build a simple radio.

Q1 Write a report of what you did.

Q2 Can you improve the performance of your radio?

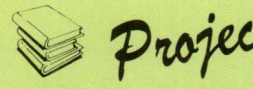

Project

Find out which part of the electromagnetic spectrum is used when you change channels on the television by remote control.

Problem

Mixing light

Splitting white light into its component colours is called **dispersion**.

What happens when we combine them together again?

Red, blue and green are **primary** colours. Colours which are made by mixing primary colours are called **secondary** colours.

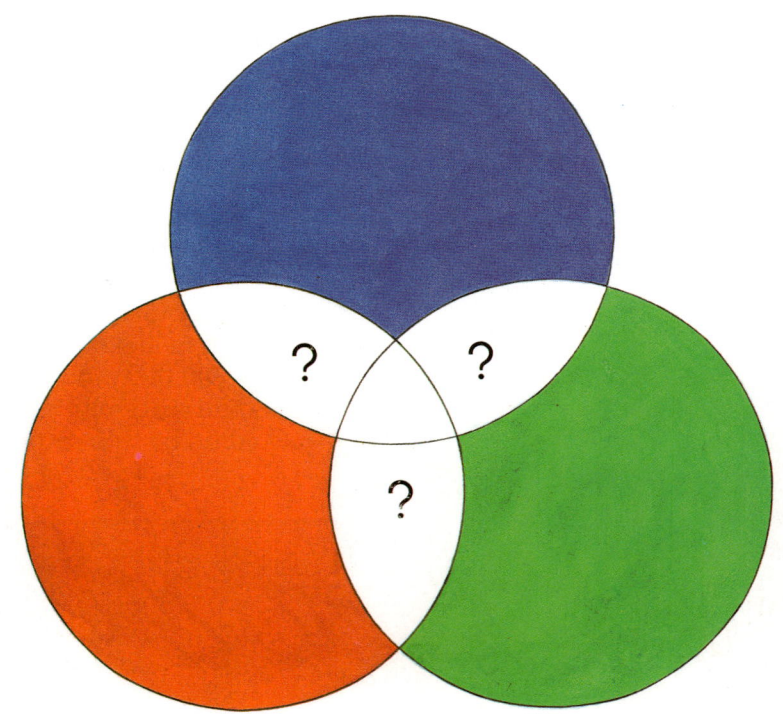

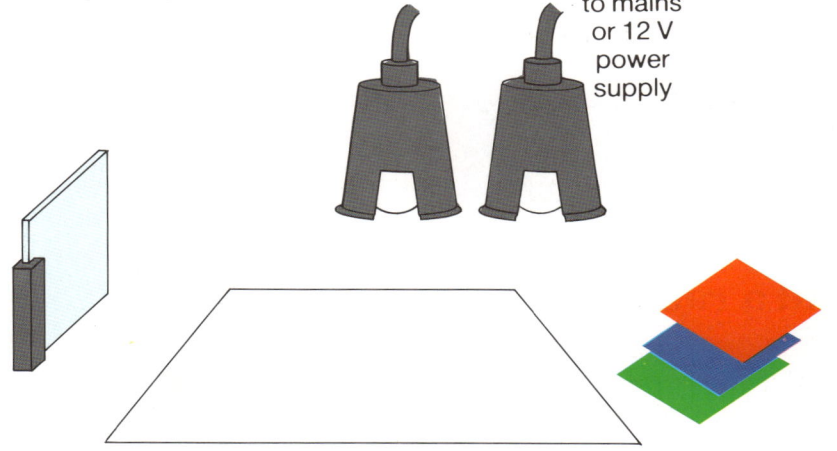

Collect
- screen
- 2 plane mirrors
- 3 coloured filters
- ray pots
- lens
- paints

You can only use two mirrors and three filters.

to mains or 12 V power supply

1 Find out what the secondary colours are.
2 What happens when you mix secondary colours?
3 Can you repeat your results with paint?

Q1 Write a report of your investigation.

Q2 Look closely at the spots on a colour television. What do you notice?

Q3 Make a poster of the television companies' logos. What do you notice about the colours they use?

Electromagnetic radiation and medicine

Work in a small group. Choose **two** of the pictures and find out more about what is happening.

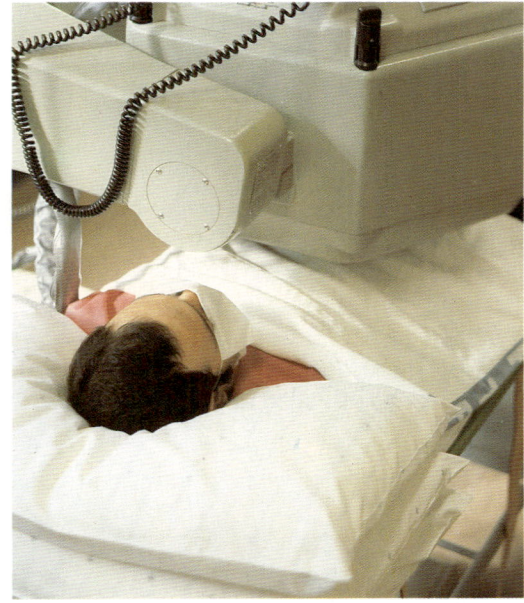

Isotope scan of a kidney, using gamma rays

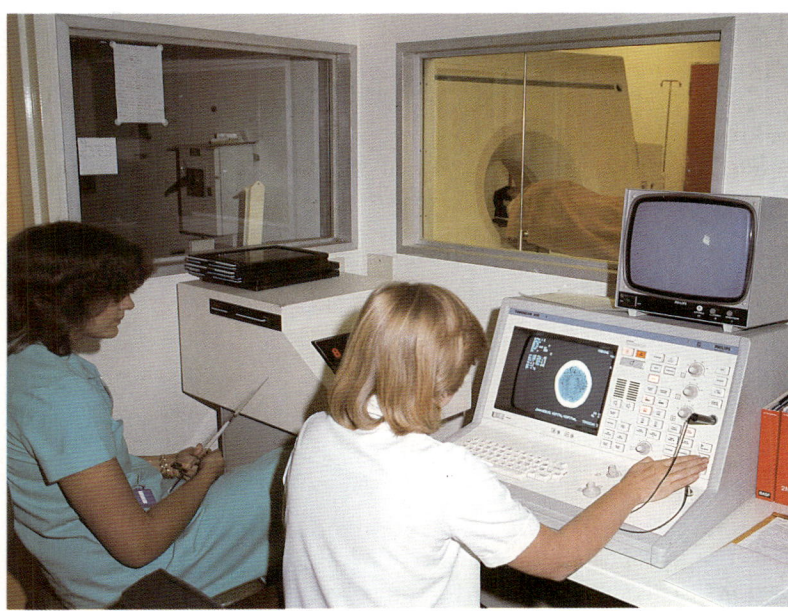

Controlling X-ray exposure to produce diagnostic images

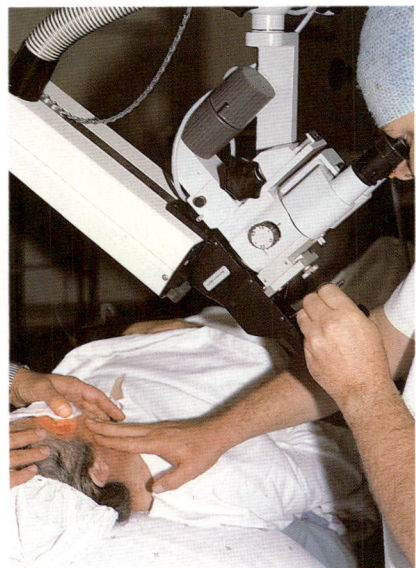

Laser surgery

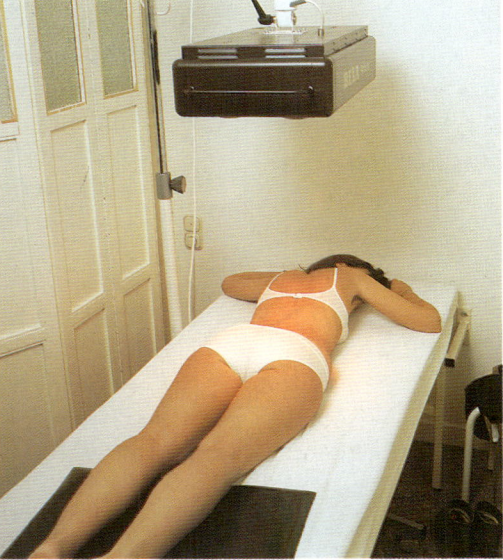

Heat treatment for muscular complaints

Prepare a short talk on each about how electromagnetic radiation is being used, the benefits and the problems.

Present your talk to the rest of the class.

1 A photographer wishes to take pictures without being noticed. He attaches two plane mirrors to his camera.

Which arrangement of mirrors will allow the photographer to take pictures of someone behind the camera? [1]

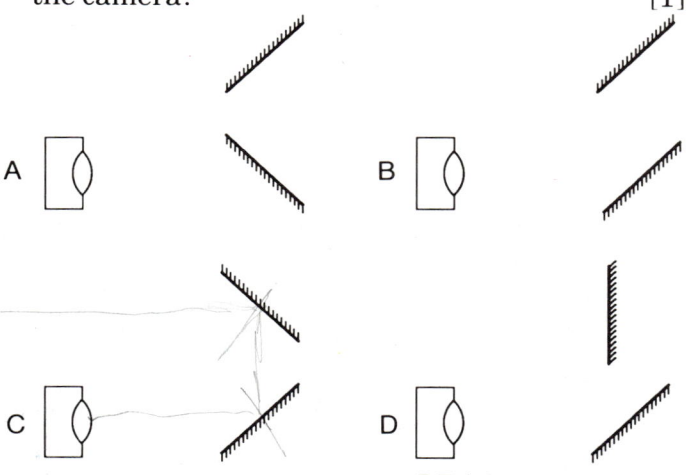

LEAG 1988 Physics

2 A model car, driven by an electric motor, is used to make a model of refraction. The car is placed on a laboratory bench and runs towards an area of sand. The car travels slowly through the sand.

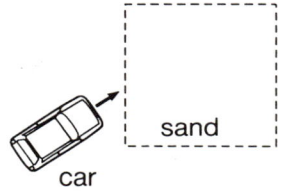

Which diagram shows the most likely path of the car? [1]

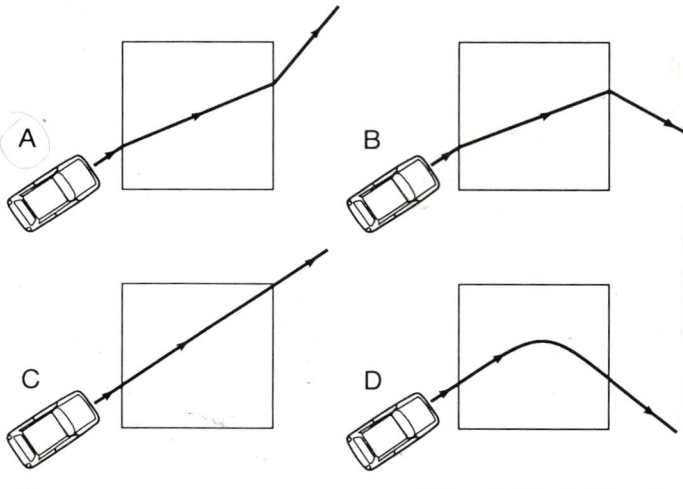

LEAG 1988 Physics

3 a The figure shows a long block of glass over an object O. Light from O reaches the top surface of the glass at X, Y and Z.

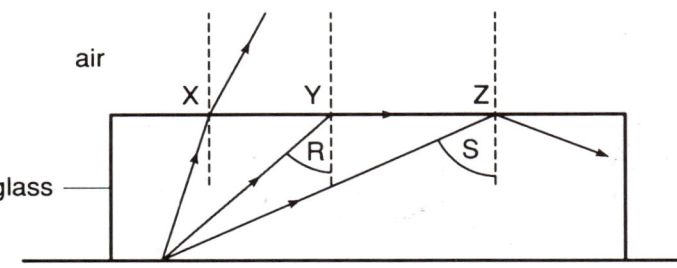

   i What is the name given to the bending of the light at X? refraction [1]
   ii Fill in the two missing words in the following sentence.
   At Z light is ...totally... ...internally... reflected. [1]
   iii What is the angle marked R called? critical [1]
   iv Why is light reflected as shown at Z? [2]
   ray is more than critical angle

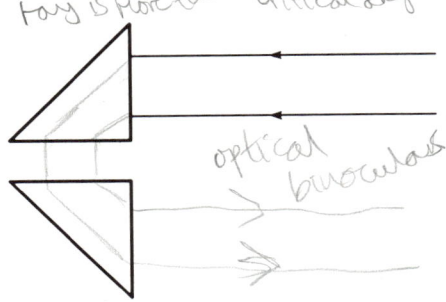

optical binoculars

b The figure above shows two 45° 45° 90° glass prisms with two rays of light incident on a face of one of them.
   i Make a sketch and complete the path of both rays through both prisms. [6]
   ii Give a practical use for such a device. [2]

MEG 1988 Physics

4 Radio Jupiter broadcasts in the Medium Wave Band on 250 m and 1200 kHz.
   a What is the *frequency* of these radio waves? [1]
   b What is the *speed* of these radio waves? Show your working. [2]
Radio Neptune transmits radio waves which have a wavelength six times larger than those of Radio Jupiter.
   c What is the *frequency* of the radio waves broadcast by Radio Neptune? [1]
   d What is the *speed* of the radio waves from Radio Neptune? [1]

MEG 1989 Physics

3
Making new materials

Mineral deposits

The mining which takes place to provide the minerals for new materials often happens in very beautiful areas. In 1949 the National Parks Commission (now the Countryside Commission) decided that ten areas of Britain should become National Parks. In these areas special care is taken to protect and enhance the landscape, and to provide opportunities for people to enjoy and appreciate them. Many people live within easy driving distance of a National Park and the Parks have become popular tourist areas.

The first National Park to be set up was the Peak National Park. The scenery is spectacular, ranging from high crags to open moorland, from limestone heights to acid peatland. Although the main industry in the area is farming, the varied geology means that a wide variety of mining also takes place.

The rocks in The Peak District were formed millions of years ago when this part of Britain was covered by sea. The two main kinds of rock to be found are **limestone** and **millstone grit**. The limestone comes from the remains of sea creatures, and fossils can often be found. Limestone is essential in many industries including the chemical industry, agriculture, building and transport, and in the production of iron and steel. Millstone grit is a hard, rough sandstone and its name comes from its historical use for grinding stones in mills. The valleys of the Peak District provided ideal sites for water-powered mills. The millstone is the symbol of the Peak National Park and it can be seen at major roads marking the boundaries of the park.

Weathered surface of limestone revealing fossils

The millstone

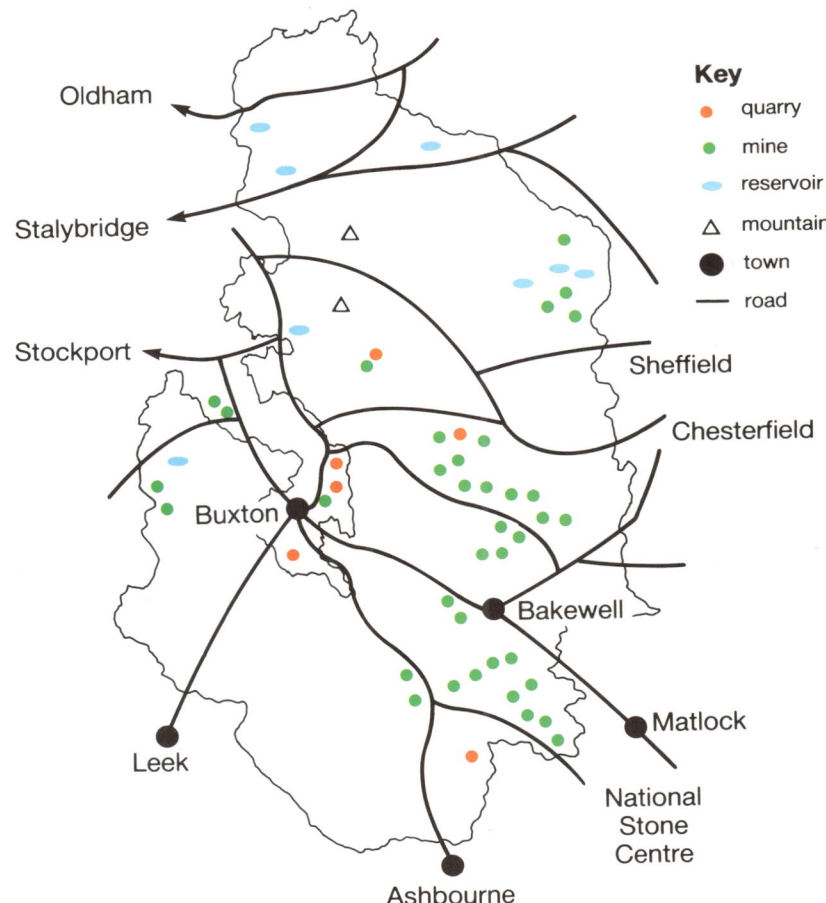

Galena

Blue John

Malachite

Lead ores such as **galena** have been mined in the Peak District since Roman times and possibly earlier. A copper ore, **malachite**, and a rare form of calcium fluoride called **Blue John** are mined for their decorative uses including jewellery.

Unfortunately mining can spoil the beauty of an area and a balance needs to be struck between national and international need for minerals as well as the local need for employment, and the need to protect the natural environment for present and future generations.

Minerals are found in the rocks of the Earth and are naturally occurring compounds of the chemical elements. If one of the elements is a metal then the mineral is known as an **ore**. The ore can be extracted from the ground and the metal separated chemically from the other elements. Some metals, such as lead, are easier to separate from their ores than others, which is why lead ore was mined as early as during the Roman occupation of Britain.

Very unreactive metals like silver and gold can be extracted from the ground as the element. They are called **native metals**. There are no deposits of native metals in the Peak District large enough to make mining economical.

Metals have many useful properties. They are good conductors of electricity and heat. They can be made into thin wires and they can be hammered into different shapes without breaking.

 *Take notes*

Discuss with your partner and write down the ways in which National Parks:

- protect habitats
- provide recreational opportunities
- support agriculture and industry.

Remember to always write a heading for your notes

Draw up a table to show the advantages and disadvantages of industry in National Parks.

Identifying minerals

Scientists have to be able to identify useful minerals, and whether, for example, a metal can be extracted from a mineral.

The two parts of a metal compound are electrically charged **ions**. The metal forms the **cation** (positively charged ion) and the other part of the compound forms the negative ion (**anion**).

$$Ca^{2+}$$
positive calcium ion

$$CO_3^{2-}$$
negative carbonate ion

A small amount of a compound held in a flame may change the colour of the flame. Testing different compounds in this way produces a pattern of results, which is shown for four metal cations in the table.

Metal cation present	Colour of flame
potassium	lilac
calcium	brick red
sodium	yellow
copper	blue/green

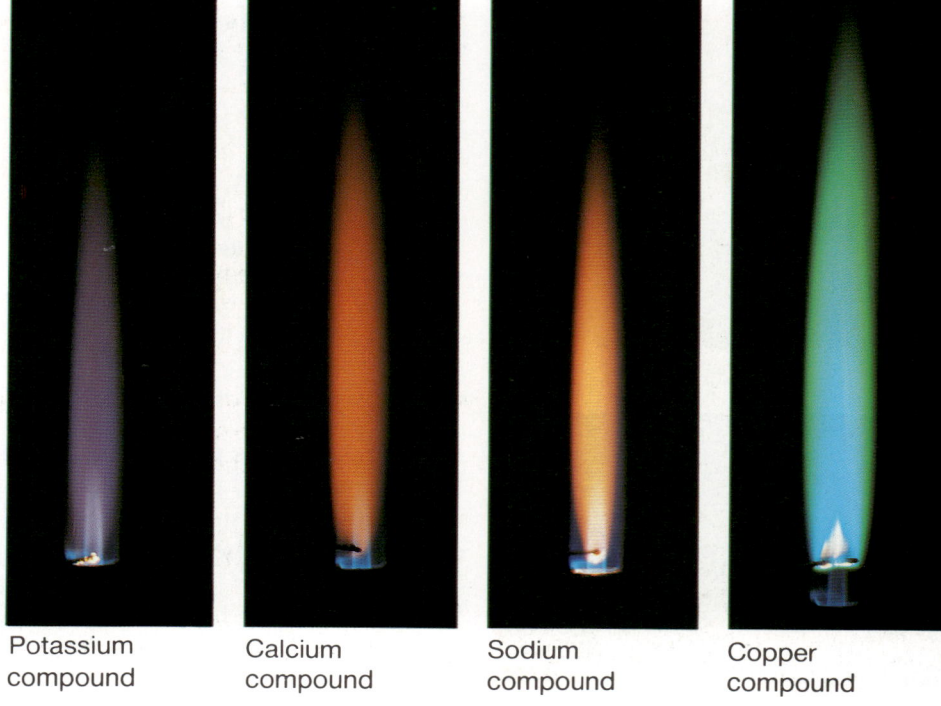

Potassium compound   Calcium compound   Sodium compound   Copper compound

Each of the metal cations burns with a characteristically coloured flame

Collect

- resource card
- named compounds
- silver nitrate solution
- dilute hydrochloric acid
- dilute nitric acid
- barium chloride solution
- test tubes and rack
- distilled water
- safety glasses

Devise a test for the negative ion by finding a pattern in your results

1 Test each of the named compounds in the three ways described on the resource card.
2 Make a table to record what happens.
3 Use your observations to design a test for each negative ion.

Q1 Describe your experiment.

Q2 Make a key to show how you can test for negative ions. You could start like this:

Bubbles of gas produced ← | Add dilute hydrochloric acid to the substance | → No bubbles of gas produced

Q3 Find metals in the Periodic Table. They are found in different sections. What are the names of the different sections?

The chemical composition of malachite

Malachite is one of the minerals that is found in the Peak District National Park. The green colour makes it particularly distinctive.

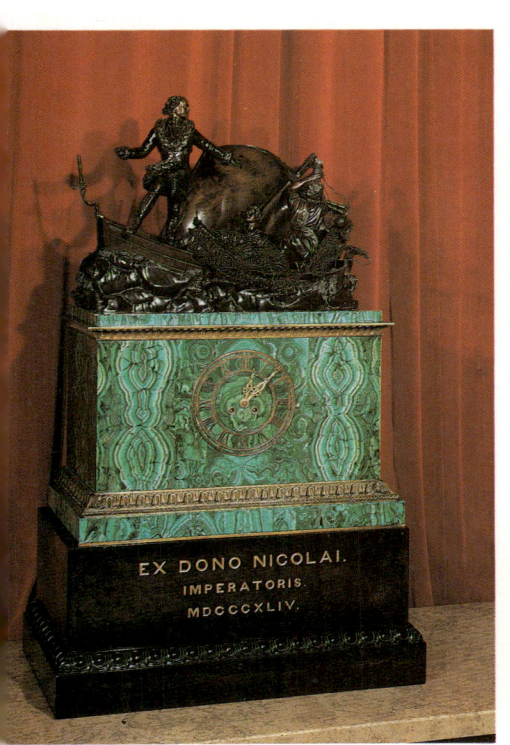

Collect

- resource cards
- powdered malachite sample
- flame-test equipment
- negative-ion test equipment
- safety glasses

Find out the chemical name of malachite

1 Carry out a flame test.
2 Carry out tests for negative ions.

Q1 Describe what you did and record all your observations. (You may want to use a table.) Remember to include your conclusion about the chemical name for malachite.

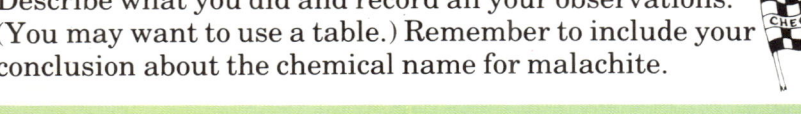
Project

Another compound that can be found in the Peak National Park is calcium carbonate. Find out how this is used in making cement.

What are acids?

You are familiar with the term **acid** and you have identified many everyday solutions as being acidic, e.g. vinegar, lemon juice. **Acidic solutions have a pH value of less than 7**.

The following demonstration will show you how an acid may be formed.

Your teacher will make the compound sulphur dioxide (SO_2) by joining the element sulphur (S) with oxygen (O_2). Water will then be added to the sulphur dioxide.

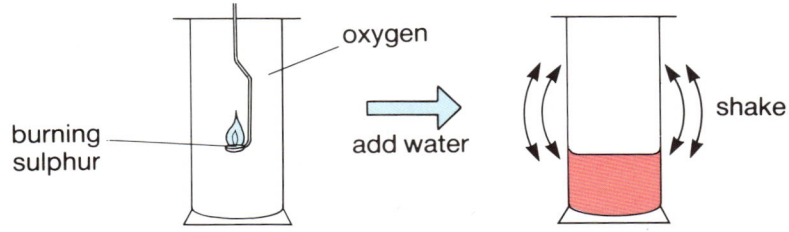

What happens when universal indicator solution is added to the solution?

Now you can try a similar experiment using carbon instead of sulphur, and a boiling tube instead of a gas jar.

Collect

- carbon sample
- tongs
- oxygen in a stoppered boiling tube in a rack
- bunsen burner and heat-proof mat
- universal indicator solution
- safety glasses

Making carbonic acid

1 Heat the carbon strongly until it is red hot.
2 Quickly unstopper the boiling tube of oxygen and plunge the carbon into the oxygen.
3 Remove the carbon when it has finished burning.
4 Add about 1 cm^3 of water to the tube, re-stopper and shake.
5 Add a few drops of universal indicator solution.

pH indicator colours

Q1 Write a description of each step of the experiment, using coloured diagrams.

Q2 Find the position of sulphur and carbon in the Periodic Table. Does this mean they are metals or non-metals?

Q3 What are the substances formed when sulphur and carbon react with oxygen? (Try writing word equations for each reaction.)

Q4 Write down the pH of the resulting solutions and say what type of substance is formed.

Q5 Which is the stronger acid, sulphuric or carbonic?

Q6 What is the reason why oxides of non-metals are called **acidic oxides**?

Properties of acids

Acids all show the same pattern of behaviour. We say that they have certain **properties** in common. Solutions of acids:

- react with many metals to form hydrogen and a salt
- react with carbonates to give carbon dioxide and a salt
- conduct electricity.

Acids can be classified as **weak** or **strong**. Strong acids have a low pH. Weak acids have a pH which is nearer to 7. Generally, weak acids come from plants and animals, e.g. vinegar can be made from grapes (it is wine that has gone sour). The acid in vinegar is called **ethanoic acid**. **Methanoic acid** comes from ants. It is this acid that ants inject into you in minute quantities when they bite. It causes a local itching sensation.

An acid can be **neutralised** by adding an **alkali** to it until its pH rises to 7. We can use this information to compare the strengths of acids.

Collect

- burette
- pipette
- conical flask
- beaker
- white tile
- lemon juice
- orange juice
- sodium hydroxide solution
- phenolphthalein indicator
- filter funnel
- clamp stand
- safety glasses

Which is the stronger acid—lemon juice or orange juice?

1 Set up the experiment as shown.
2 Note the volume of sodium hydroxide in the burette.
3 Add a few drops of indicator to the flask.
4 Add the sodium hydroxide slowly and shake the flask gently until the indicator just turns red.
5 Repeat using orange juice.

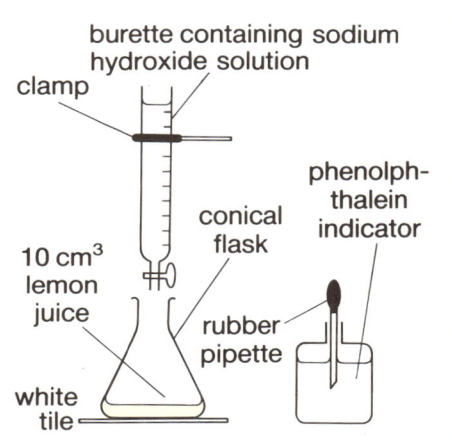

Q1 Write an account of the experiment. Present your results in a table like the one on the right.

Q2 Which juice was the more acidic?

Volume of sodium hydroxide (cm³)	Lemon juice	Orange juice
Second burette reading		
First burette reading		
Volume of sodium hydroxide needed to neutralise		

The manufacture of sulphuric acid

Sulphuric acid is a very important chemical as it is used to make many of the goods we buy.

Runcorn in Cheshire has one of many factories in the UK which manufactures sulphuric acid. The raw materials needed are oxygen, which can be obtained from air, and sulphur which is imported from the USA. Other manufacturing plants import sulphur from Poland.

The process used for sulphuric acid production is called the **Contact Process**. There are three stages.

1 Furnace

Sulphur dioxide is produced by burning a fine spray of liquid sulphur in air.

$$\text{sulphur} + \text{oxygen} \rightarrow \text{sulphur dioxide}$$
$$S(s) + O_2(g) \rightarrow SO_2(g)$$

2 Converter

Sulphur dioxide is mixed with more air and passed through several layers of **catalyst** at 450 °C. The catalyst used is vanadium V oxide. This produces sulphur trioxide. Remember that a catalyst speeds up a reaction without undergoing a permanent change itself.

$$\text{sulphur dioxide} + \text{oxygen} \rightleftharpoons \text{sulphur trioxide}$$
$$2SO_2(g) + O_2(g) \rightleftharpoons 2SO_3(g)$$

The reaction is **reversible**, which means that not all of the sulphur dioxide becomes sulphur trioxide. The sulphur trioxide can split up into sulphur dioxide and oxygen again. Conditions are chosen to make the oxygen and sulphur dioxide combine as quickly as possible compared with the reverse reaction. A very high temperature would damage the catalyst. 450 °C is used as this produces the fastest rate of reaction without decomposing the catalyst. Removing the sulphur trioxide also helps to prevent it breaking up into sulphur dioxide and oxygen.

Use of sulphuric acid	Percentage
Fertiliser/other agricultural use	30.4
Chemicals	16.4
Paints/pigments	15.7
Detergents/soaps	11.7
Natural and synthetic fibres	9.4
Metal working	2.3
Dyes	2.0
Oil/petrol	1.0
Other uses	11.1

Any sulphur dioxide or sulphur trioxide which escapes into the air will produce acid rain which results in damage to the natural environment such as forests and lakes and also to the fabric of limestone buildings

3 Absorber

The sulphur trioxide could in theory be dissolved in water to produce sulphuric acid. However, this reaction gives out so much heat that the sulphuric acid is formed as a fine mist of droplets which cannot be collected.

In order to avoid this the sulphur trioxide is absorbed in 98% sulphuric acid. This forms a fuming liquid called **oleum**. This can then be diluted to sulphuric acid of the desired concentration.

1 Collect jigsaw pieces and complete the jigsaw of a sulphuric acid plant.

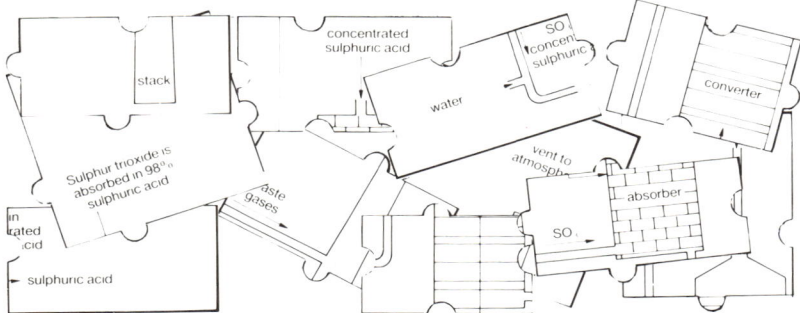

2 Draw a flow diagram to show the stages in the manufacture of sulphuric acid.

Q1 Why do you think sulphur is transported in its liquid state?

Q2 During the Contact Process, hot sulphur dioxide is cooled by passing through a heat exchanger. The heat turns water into steam which can be used to generate electricity. Do you think generating electricity in this way is a good idea?

Q3 Why must expensive filters be installed to remove any trace of acid spray from the waste gases as they escape through a chimney?

Q4 Besides the cost of raw materials, what other hidden costs must be covered by the manufacturer?

Q5 Find Runcorn on a map of the United Kingdom. List all the reasons why you think a sulphuric acid plant was built there.

Q6 Show the information about the uses of sulphuric acid as a pie chart.

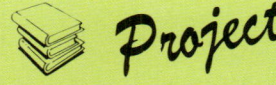

Project

Find out why vinegar can remove limescale from kettles.

Bases and alkalis

What are bases and alkalis?

A **base** is any substance that can neutralise an acid. Oxides and hydroxides of metals are bases. Most bases are insoluble in water. Metal hydroxide bases can be **precipitated** when sodium hydroxide solution is reacted with a solution of a metal compound, e.g.

copper chloride + sodium hydroxide → copper hydroxide (base) + sodium chloride

$$CuCl_2(aq) \quad + \quad 2NaOH(aq) \quad \rightarrow \quad Cu(OH)_2(s) \quad + \quad 2NaCl(aq)$$

The insoluble copper hydroxide will be seen as a **precipitate** (solid).

You can make some metal hydroxides (bases) by carrying out the following activity.

Collect
- 4 test tubes and rack
- sodium hydroxide solution
- copper sulphate solution
- iron (II) sulphate solution
- iron (III) chloride solution
- aluminium chloride solution
- safety glasses

1 Put 2 cm³ copper sulphate solution into a test tube.
2 Add drops of sodium hydroxide solution until you see a change.
3 Repeat using the other solutions.

Q1 Describe how you did the experiment.
Copy and complete the following results table:

Solution	Observation after addition of sodium hydroxide	Name of base produced

Q2 Try to write word equations for the reactions.

An **alkali** is a metal hydroxide dissolved in water.

Your teacher will burn sodium and calcium in air. Each metal will react with the oxygen in the air to form sodium oxide or calcium oxide. These oxides are bases and they dissolve in water to produce sodium hydroxide and calcium hydroxide, which are both alkalis. **An alkali is formed when a soluble base dissolves in water.** Remember, most bases are insoluble.

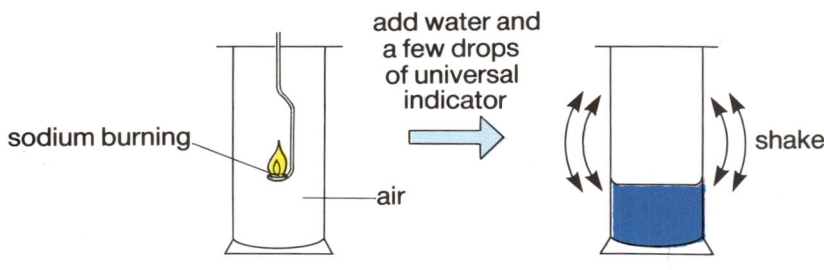

sodium burning

air

add water and a few drops of universal indicator

shake

Q3 Describe the demonstration. Copy the table below and complete it.

Element	Observation on burning	Name of product	Name of solution	Colour of universal indicator in the solution

The chemical reactions for sodium can be written as follows:

sodium + oxygen → sodium oxide (base)

$$4Na(s) + O_2(g) \rightarrow 2Na_2O(s)$$

sodium oxide + water → sodium hydroxide (alkali)

$$Na_2O(s) + H_2O(l) \rightarrow 2NaOH(aq)$$

The common alkaline solutions found in the laboratory are:

- sodium hydroxide solution
- potassium hydroxide solution
- calcium hydroxide solution
- barium hydroxide solution.

Consider the ions that are present in alkaline solutions:

Solution	Ions present
sodium hydroxide	sodium Na^+ hydroxide OH^-
potassium hydroxide	potassium K^+ hydroxide OH^-
calcium hydroxide	calcium Ca^{2+} hydroxide OH^-
barium hydroxide	barium Ba^{2+} hydroxide OH^-

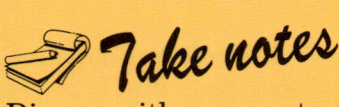

Take notes

Discuss with your partner and write down:

- why metal oxides and hydroxides are bases
- how alkalis are formed
- the ion present in potassium hydroxide solution that makes it an alkali.

CHECKPOINT

There is a pattern. All the alkalis produce hydroxide ions when they are dissolved in water. It is the hydroxide ion that gives them their common alkaline properties.

Slaked lime

We have already mentioned the presence of limestone in the Peak National Park. About 20% of limestone quarried in the UK is converted into lime and then into slaked lime by adding the lime to water.

Slaked lime is used
- for neutralising acid soil
- in the manufacture of bleaching powder
- for softening temporarily hard water
- in the manufacture of paper.

Slaked lime being spread on acid soil

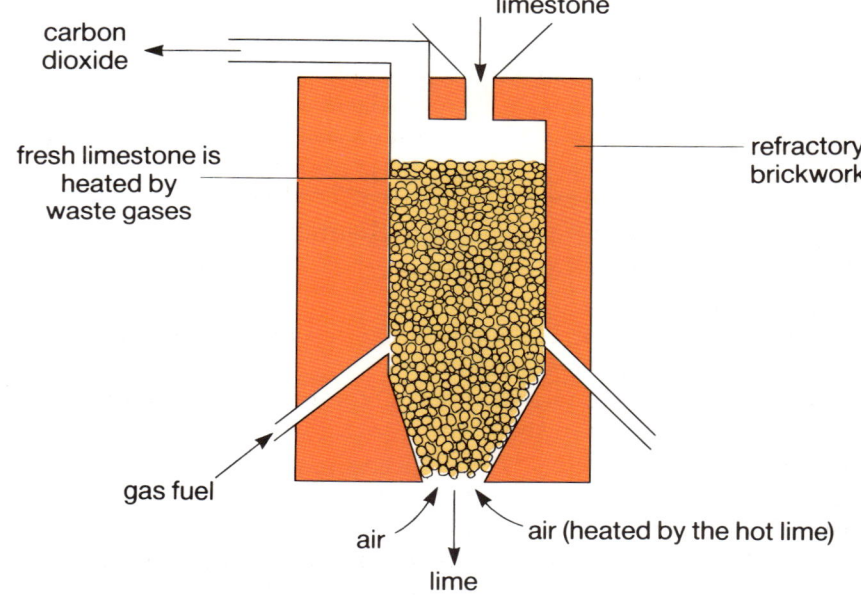
The lime is produced in a lime kiln

Limestone (calcium carbonate) is converted into lime (calcium oxide) by heating it to about 1000°C. This is an example of **thermal decomposition** (breaking down a substance using heat).

$$limestone \rightarrow lime + carbon\ dioxide$$
$$CaCO_3(s) \rightarrow CaO(s) + CO_2(s)(g)$$

The lime can be mixed with water to form slaked lime (calcium hydroxide).

$$lime + water \rightarrow slaked\ lime$$
$$calcium\ oxide + water \rightarrow calcium\ hydroxide$$
$$CaO(s) + H_2O(l) \rightarrow Ca(OH)_2(s)$$

 Take notes

Discuss with a partner, then:

- draw a flow diagram to show how slaked lime is produced in a lime kiln
- write down where the limestone for the process might be quarried
- write down why slaked lime is important to farmers
- write down three other uses of slaked lime.

1 Place 1 cm³ of each salt solution into separate test tubes.
2 Add silver nitrate solution to each.
3 Repeat **1** but this time add barium chloride solution.

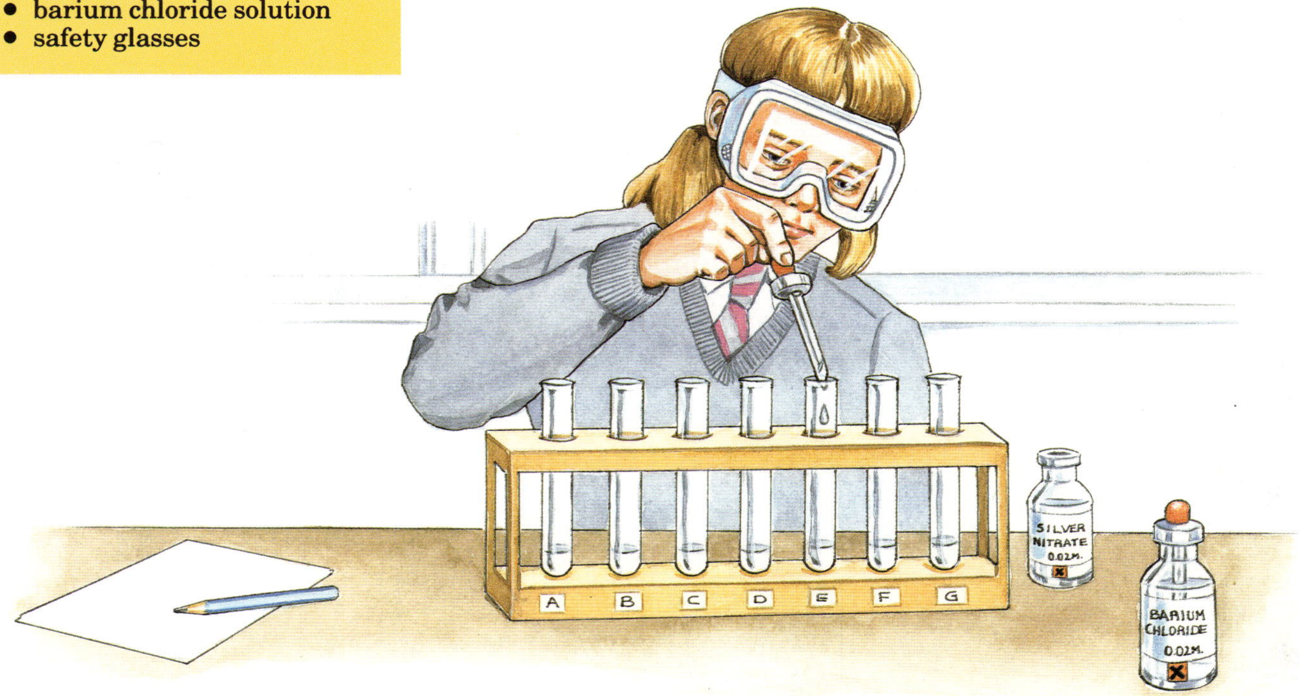

Q1 Describe how you did the experiment.

Q2 Record your observations.

Q3 Write a word equation for the reactions that produced insoluble salts.

Take notes

Discuss in your group and write down

- how a salt is named
- how a salt is formed
- the four general equations for making soluble salts
- a method for making insoluble salts
- word equations for preparing:
  a copper sulphate from copper carbonate
  b zinc nitrate from zinc oxide
  c magnesium chloride from magnesium
  d potassium sulphate from potassium hydroxide
  e lead sulphate from lead nitrate.

Project

Find out using the library, or any other sources you think may be helpful, the origin of the expressions *worth your salt* and *sitting below the salt* and the origin of the word *salary*.

Speeding up

Controlling the rate of a chemical reaction

Chemical reactions take place at different speeds. Some reactions are very fast, e.g. explosions. Some take place at moderate speeds, e.g. metals dissolving in acids. Others are very slow, e.g. the rusting of iron.

It is important to know what the speed of a reaction is and how we can alter the speed. If, for example, the reactions that take place when manufacturing sulphuric acid are too slow then the process might be uneconomical.

Collect

- measuring cylinder
- conical flask
- magnesium ribbon (4 cm)
- dilute hydrochloric acid
- clock
- syringe or trough and graduated glass tube
- glass tubing
- safety glasses

Investigate reaction rate

1 Put 40 cm^3 acid into a conical flask.
2 Set up the apparatus as shown.
3 Drop the magnesium ribbon into the flask and quickly replace the bung.
4 Read the volume of gas every ten seconds until it remains constant.
5 Make a table of your results:

Time (s)	Volume of gas (cm^3)

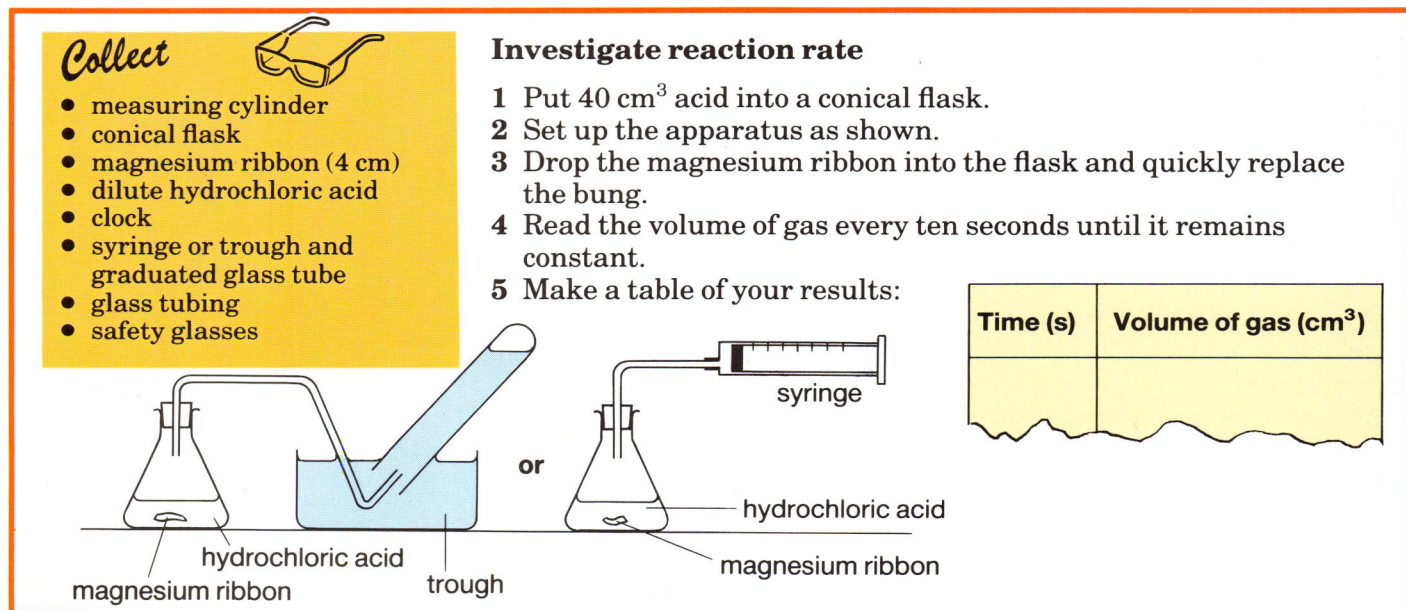

hydrochloric acid

magnesium ribbon   trough   **or**   hydrochloric acid

magnesium ribbon

syringe

Q1 Describe how you did the experiment.

Q2 What is the name of the gas that you collected?

Q3 Plot a graph of volume of gas against time.
Look at your graph and answer the following questions.

  a At what time did the reaction end? How do you know?
  b How many cm^3 of gas had been collected at the end of the experiment?
  c Was the speed of reaction constant or did it slow down or speed up? How do you know?
  d During what part of the experiment was the speed fastest? How do you know?

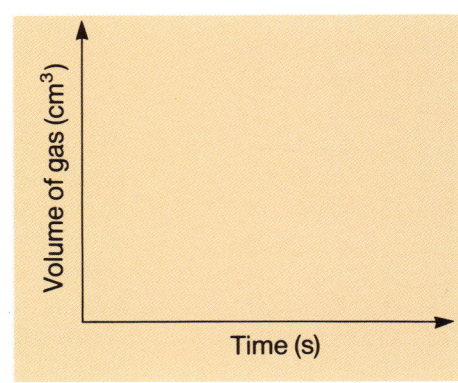

What changes the rate of reaction?

Using hydrochloric acid and magnesium ribbon, investigate the effects of **reactant particle size**, **concentration of acid**, and **heat** on the rate of reaction.

1 Design your experiment. Remember:
   - only change one variable at a time.
   - each variable should be changed at least three times to find a pattern, e.g. try three different temperatures.
2 Show the plan to your teacher and carry out the experiment.

Note When investigating the effect of heat, turn off the bunsen burner before adding the magnesium.

Q4 Describe your experiments, making clear how they were made valid.

Q5 Record your results in a table.

Q6 If you have found a pattern in your results you should be able to say how the following affect the reaction rate:
   a changing particle size
   b changing concentration
   c changing temperature.

Catalysts

You should be familiar with the term catalyst and know that a catalyst is used to speed up the rate of a chemical reaction without being chemically changed itself. For example, vanadium V oxide is used as a catalyst in the industrial manufacture of sulphuric acid.

Living organisms contain biological catalysts called **enzymes**. For example, **peroxidase** is the enzyme produced by animals and plants to break down harmful peroxides into oxygen and water.

'Biological' washing powders contain enzymes to break down fats etc.

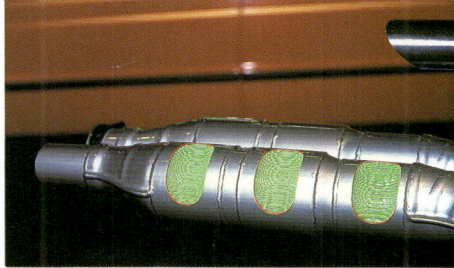

Catalytic converters promote oxidation reactions which convert harmful gases in car exhaust to less harmful ones

Collect

- test tubes and rack
- teat pipette
- hydrogen peroxide
- measuring cylinder
- freshly puréed celery
- freshly puréed potato
- freshly puréed liver
- safety glasses

Investigate a biological catalyst

1 Put 2 cm³ hydrogen peroxide into a test tube.
2 Add a full teat pipette of puréed celery.
3 Observe any change.
4 Repeat with potato.
5 Repeat with liver.

Q1 Describe what you did.

Q2 Which substance(s) contained most peroxidase? How do you know?

Take notes

Think about and write down

- the names of two reactions you know are fast
- the names of two reactions you know are slow
- why it is useful to be able to control the rate of a reaction
- four ways of speeding up the rate of a reaction
- what an enzyme is
- the name of one enzyme produced by both animals and plants and why it is needed.

The collision theory

Experiments have shown us that the speed of chemical reaction can be increased if you

a increase the concentration of a solution
b increase the temperature
c increase the surface area of solid particles
d use a catalyst.

In order to explain this you have to use your ideas about particles and how they behave.

For a reaction to take place the particles must collide with enough energy to break the **bonds** which hold them together. The minimum energy needed in collisions before a reaction can occur is called the **activation energy**.

The rate of reaction is increased if you increase the number of collisions that have enough energy. You can do this in four ways.

1 Increasing the concentration

Particles are more likely to collide if the concentration of the solution is greater, so there will be more frequent collisions.

2 Increasing the temperature

If the temperature is raised the particles move faster so there are likely to be more collisions in a given time. They also have more energy when they do collide so it is more likely that the bonds will be broken.

3 Increasing the surface area

Collisions take place on the surface of a solid particle. Therefore, if there is a larger total surface area there will be more frequent collisions and a faster reaction.

4 Using a catalyst

Catalysts work in a complex fashion. They provide an alternative reaction mechanism with a lower activation energy. This means that more of the collisions are successful.

Investigate change of concentration

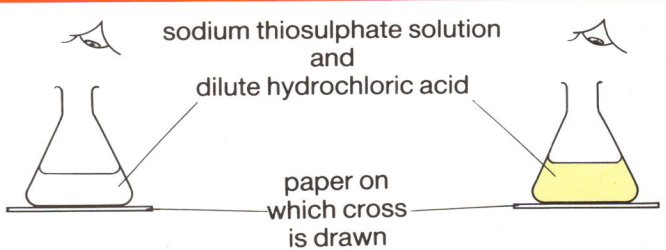

sodium thiosulphate solution and dilute hydrochloric acid

paper on which cross is drawn

START—cross can be seen

END—cross cannot be seen

Follow the instructions on the experiment sheet carefully and record your observations.

Sodium thiosulphate solution reacts with dilute hydrochloric acid to produce a precipitate of yellow sulphur.

sodium thiosulphate + hydrochloric acid → sodium chloride + water + sulphur dioxide + sulphur

$Na_2S_2O_3(aq) + 2HCl(aq) \rightarrow 2NaCl(aq) + H_2O(l) + SO_2(aq) + S(s)$

The rate of production of sulphur is a measure of the rate of reaction.

Q1 Describe the experiment and record your results in a table.

Q2 Plot a graph with concentration of sodium thiosulphate on the vertical axis and time on the horizontal axis.

Q3 Do you think your results were accurate? How do you know?

Research project

- Find out which are the main manufacturing industries in your area.

- Identify the raw materials they use and find out where they come from. Do any of the industries involve extracting minerals? How are the materials transported?

- What environmental effects does each industry have on the region?

- What steps can be taken to reduce any harmful effects on the environment?

- In what ways does your area benefit from these industries?

Work in a group and plan a report on your research findings. You can present your report in any of the ways illustrated here.

Friendly products

Compare the information given on the different products shown.

The top picture shows products which have recently appeared on the supermarket shelves. Suggest reasons why they have been developed.

1 Limestone is a naturally occurring form of calcium carbonate. It is an important raw material used in many different industries.

a Explain why the quarrying of limestone can cause environmental problems. [1]

b One of the properties of limestone is its reaction with acids.
i How can buildings made of limestone be damaged by atmospheric pollution? [1]
ii Why do some farmers put powdered limestone on their fields? [1]
iii Write a word equation for the reaction of calcium carbonate with nitric acid. [1]

c Name a building material which is made by heating a mixture of limestone and clay. [1]

d Name one other industrial process in which limestone is used. [1]

NEA 1985 Chemistry (II)

2 Four bottles of dilute salt solutions in a school laboratory had lost their labels. They were known to contain iron(II) sulphate, iron(III) chloride, copper(II) sulphate and calcium chloride.

Describe how a student could use sodium hydroxide solution to identify the substances in the bottles. State any important observations and deductions. [4]

NEA 1989 (Nov.) Chemistry Syllabus A (II)

3 Epsom salts (magnesium sulphate), calcite (calcium carbonate) and halite (sodium chloride) are all colourless minerals.

Give the details of **three** chemical tests which a geologist could use to identify the sulphate, the carbonate and the chloride. State the result of each test. [8]

Write an equation for **one** reaction occurring in any of the tests. [1]

NEA 1989 Chemistry Syllabus A (II)

4 Silver(I) bromide is prepared by mixing solutions of silver(I) nitrate and potassium bromide. The cream-coloured precipitate of silver(I) bromide is filtered off, washed and spread over the filter paper. A coin is placed on this coated filter paper and a piece of burning magnesium ribbon is held above it.

This experiment illustrates the photochemical decomposition of silver(I) bromide into its elements.

i Write a word for decomposition of silver(I) bromide. [1]

ii Describe what you would observe at the end of the experiment when the coin is removed. [2]

iii Explain the observations you have given in **ii**. [1]

iv Name one application of this decomposition of silver(I) bromide. [1]

NEA 1984 Chemistry (II)

5 In an experiment to investigate the rate of a reaction a sample of calcium was reacted with an excess of water at room temperature ($20\,^\circ C$) and atmospheric pressure. Hydrogen was given off and the volumes obtained at different time intervals are shown in the following table.

Time (minutes)	1	2	3	4	5	6
Volume of hydrogen (cm^3)	29	44	53	59	60	60

i Draw a diagram of an apparatus suitable for carrying out this experiment. [2]

ii Plot the results on graph paper. Label this graph **X**. [2]

iii Explain why the first part of the graph has the steepest gradient. [1]

iv Explain why the last part of the graph is a horizontal line. [1]

v After what time was the reaction half completed? [1]

vi Sketch on the same axes the graphs you would obtain if **only** the following changes were made in the original conditions.
a The reaction temperature was $40\,^\circ C$. Label this graph **Y**. [2]
b Half the original mass of calcium was used. Label this graph **Z**. [2]

NEA 1985 Chemistry (II)

4

Moving electrons

Charging around

Thales of Miletus was a Greek philosopher around 600BC who is thought to have rubbed a piece of amber on his sleeve to clean it. He dropped it accidentally in some dried leaves and when he picked it up he found, to his surprise, that some leaves were clinging to the amber.

As Thales rubbed the amber against his sleeve, small particles carrying negative charge moved from his sleeve to the amber, leaving the amber negatively charged. The leaves were attracted by the negative charge on the amber.

These negatively charged particles are the tiny **subatomic** particles which orbit the nucleus of an atom—**electrons**. The nucleus contains **protons** which are positively charged, as well as **neutrons** which have no charge. Normally the negative and positive charges in an atom cancel one another so normal matter composed of **neutral** atoms is not charged.

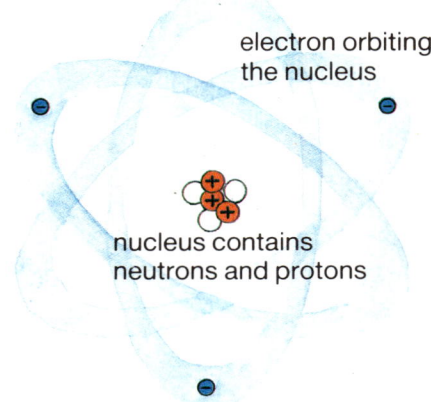

electron orbiting the nucleus

nucleus contains neutrons and protons

The nuclear atom

The transfer of negatively charged **electrons** by rubbing had upset the charge balance, and the resulting attraction was an effect of **static electricity**.

You should be familiar with attractive and repellant forces between electrically charged objects. If not, look up the topic ('electric charge') in a science book like *Understanding Science 3*.

Current electricity

Current electricity is the movement of charged particles like electrons through a material. **Current** is the **rate of flow of charge**. It is measured in **amperes**, or **amps**.

Materials that electrons can pass through are called **conductors**. A conductor can be 'good' or 'bad'. Electrons can pass with **small resistance** through a good conductor but there is greater resistance to their motion in a poor conductor. Materials with very **high resistance** are called **insulators**.

To make electrons move in a conductor there must be a source of energy. You are probably familiar with a Van de Graaff generator (if not, look it up). When this machine is 'charged up' by its rotating belt, electrons can be forced to move down an attached dampened thread.

Batteries also generate a current. Compact 'dry' batteries do not have to be charged up—they have the **stored potential** to produce a current. The size of the potential is measured in **volts** and is often referred to as the 'voltage' of the battery. When a battery is connected to a conducting **circuit**, electrons flow.

The rate of flow of charge around the circuit is known as the **current**

The **resistance** of the filament wire in the lamp is high so the electrons need to expend energy to get through it. This energy is given off as light and heat

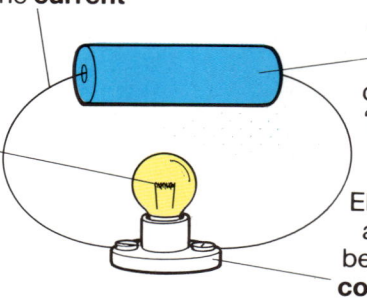

The battery 'pushes' electrons round the circuit. The amount of 'push' given is known as the **potential**

Electrons can move around the circuit because the wire is a **conductor** (it has low resistance)

Ammeters and voltmeters

An **ammeter** is used to measure the size of an electric current in amps, A. An ammeter must be connected correctly into a circuit.

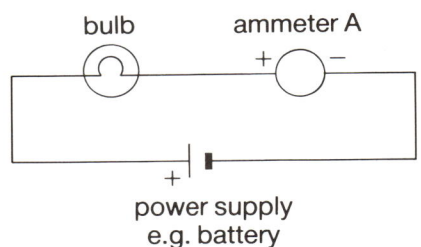

bulb   ammeter A
+   −

power supply
e.g. battery

The ammeter must be connected *in series* (in line) with the other components in the circuit. The positive end of the meter must connect with the positive end of the power supply

A **voltmeter** is used to measure the **potential difference** between two points of a circuit. Remember, a battery or other power supply has electric potential, and this is gradually used up round the circuit in 'pushing' the electrons. Consequently there is a drop in potential between any two points of a circuit which depends on the resistance (the amount of 'pushing' required) between those points. Potential difference is measured in volts, V.

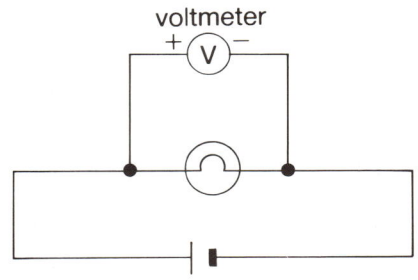

voltmeter
+ V −

The voltmeter must be connected *in parallel* with (across) the part of the circuit whose potential difference is being measured. The positive end of the meter must connect with the positive end of the power supply

Measuring current and potential difference

Work with a partner and connect up a circuit as shown. Remember that the positive terminal of each meter must connect with the positive terminal of the power supply.

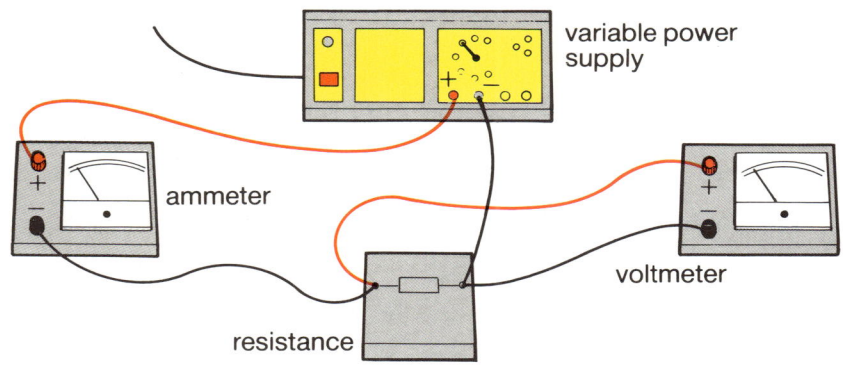

variable power supply

ammeter

voltmeter

resistance

1 Increase the voltage from the power supply in steps of 1 V.
2 Each time, read the ammeter and the voltmeter and note the readings.

Work out the values of divisions and subdivisions on your meter scales so you can calculate a reading correctly when the pointer falls between labelled values

To make your readings as accurate as possible, always read the meter from directly above the pointer. You and your partner should each read the meter and then compare readings

3 Make a table of your current and potential difference readings.
4 Plot a graph with potential difference on the vertical axis and current on the horizontal axis. Draw the best straight line through your plotted points and work out the **gradient** of the line.

$$\text{Gradient} = \frac{a}{b}$$

potential difference (V)

a

b

current (I)

Resistance

The gradient of your graph is the value of the **resistance** across which the potential difference was measured.

$$\text{resistance} = \frac{\text{potential difference}}{\text{current}}$$

Resistance is measured in **ohms**, Ω. In symbols, $R = \dfrac{V}{I}$.

For metallic conductors if the temperature is kept constant R is constant (hence your straight-line graph). This means that **the current is proportional to the potential difference**.

This is known as **Ohm's law**.

Collect

- power supply
- ammeter
- voltmeter
- variable resistor
- connecting leads
- filament lamp
- semiconducting diode

Investigate the resistance of a filament lamp and a semiconductor diode

1 Draw the circuit you will use. Check the drawing with your teacher and then set up the circuit using the available apparatus.
2 How many readings should you take?
3 Take the readings and tabulate them.

[IT] A spreadsheet could be used for the results.

Q1 Write an account of your investigation.

Q2 Draw a graph of the results of each experiment. What shapes are your graphs?

Q3 How does the resistance of the semiconductor diode differ from that of the filament lamp? Does it obey Ohm's law?

More resistance

What happens when we put more than one resistor in the same circuit?

When resistors are connected **in parallel** with one another the combined resistance is found by using the following formula:

$$\frac{1}{R_{\text{total}}} = \frac{1}{R_1} + \frac{1}{R_2}$$

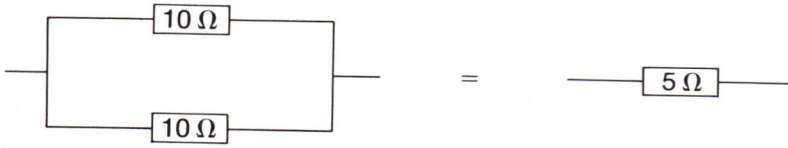

Work with a partner.

1 Set up the circuit with the two resistors in series.
2 Take measurements with the voltmeter and ammeter as before and tabulate your readings.

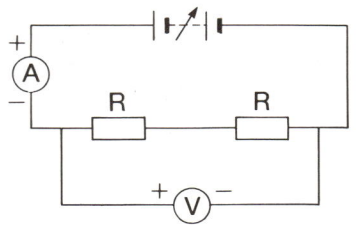

Resistors in series

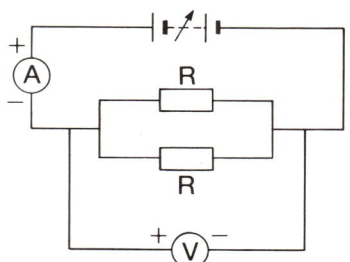

Resistors in parallel

3 Alter the circuit so the resistors are in parallel.
4 Take readings and tabulate them.

[IT] A spreadsheet could be used for the results.

5 By drawing graphs, work out the total resistance of the resistors in each case. What do you find?

Q1 How can you work out the total resistance in a circuit if the resistors are combined in series?

Q2 Work out the total resistance of your two resistors in parallel using the formula at the bottom of p. 68. Does it agree with your experimental result?

How much resistance?

Resistors are found in many everyday devices like radios and televisions. They have four coloured bands. The first two bands give the digits of the value. The third band gives the number you must multiply the digits by to get the correct value of the resistance. The fourth band is either gold or silver and this indicates how accurate the resistor is.

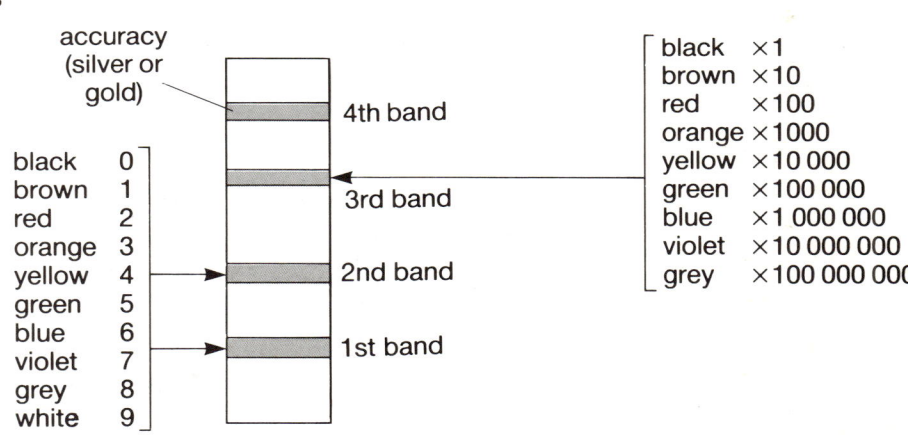

black	0
brown	1
red	2
orange	3
yellow	4
green	5
blue	6
violet	7
grey	8
white	9

accuracy (silver or gold)
4th band
3rd band
2nd band
1st band

black	×1
brown	×10
red	×100
orange	×1000
yellow	×10 000
green	×100 000
blue	×1 000 000
violet	×10 000 000
grey	×100 000 000

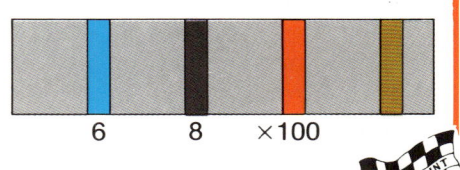

Project

Find out what a variable resistor is. What is it used for?

All electrical goods have a power 'rating' marked on them

Power

Electrical appliances turn electrical energy into other forms of energy such as heat, light, sound or movement.

The faster an appliance converts energy, the more **powerful** it is.

Energy is measured in joules (J). The **power** of a device tells you how many joules of energy it can convert to other forms of energy in every second.

$$\text{power} = \frac{\text{energy used}}{\text{time taken}}$$

The unit for power is therefore J/s which is commonly known as the watt (W). 1 kilowatt (kW) = 1000 watts (W).

Investigate the behaviour of various electrical devices

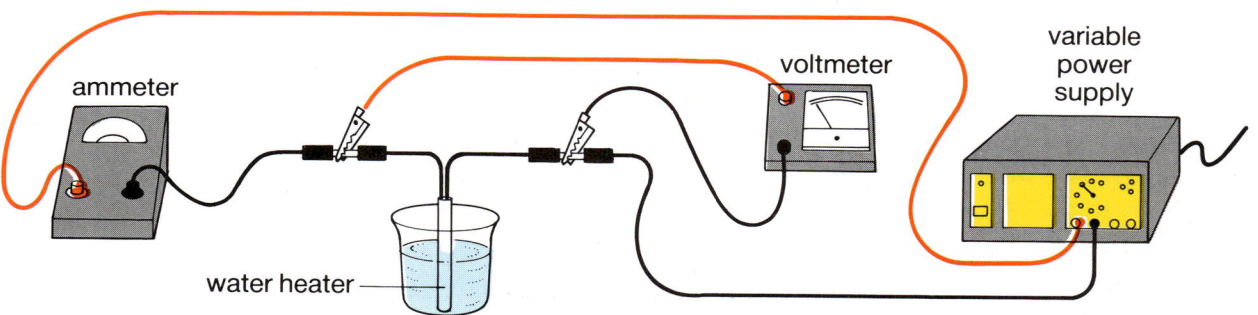

ammeter   water heater   voltmeter   variable power supply

Collect
- power supply
- voltmeter
- ammeter
- connecting leads
- water heater
- beaker of water
- light bulb
- electric motor

1 Set up the circuit as shown.
2 Adjust the power supply until the water heater is working at its recommended voltage.
3 Note the readings on the ammeter and voltmeter.
4 Repeat the experiment using a light bulb.
5 Repeat with an electric motor.

Q1 Draw circuit diagrams for your experiments. Invent symbols for the heater and the motor.

Q2 Put your results in a table as follows. Take the power value *P* from the device.

I	V	P

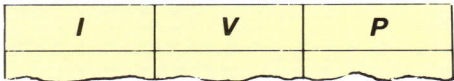 A spreadsheet could be used for the results.

Q3 Look for a simple mathematical connection between *P*, *V* and *I*. Use just the four rules ($\div$, $\times$, $+$, $-$).

Q4 In the UK the supply voltage in most homes is 240 V. Work out how much current is passing through the following household appliances when they are working.

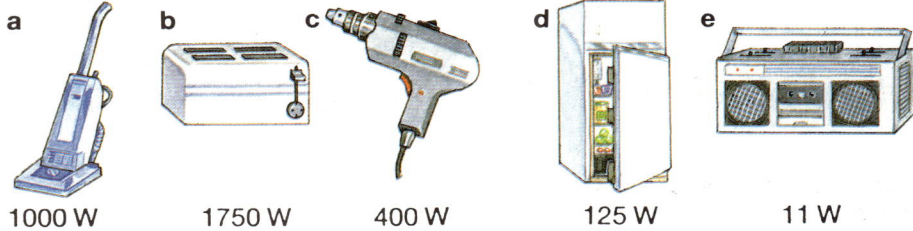

a	b	c	d	e
1000 W	1750 W	400 W	125 W	11 W

Take notes

Discuss with your partner and write down:

- what is meant by the *power* of an electrical device
- two equations which can be used to calculate the power.

CHECKPOINT

Fuses

If too large a current passes through some appliances it can damage them. To prevent damage when there are current fluctuations, plugs on appliances should be fitted with a fuse. A fuse is a short length of wire made of a metal with a low melting point. When the current through it exceeds a certain value, it melts.

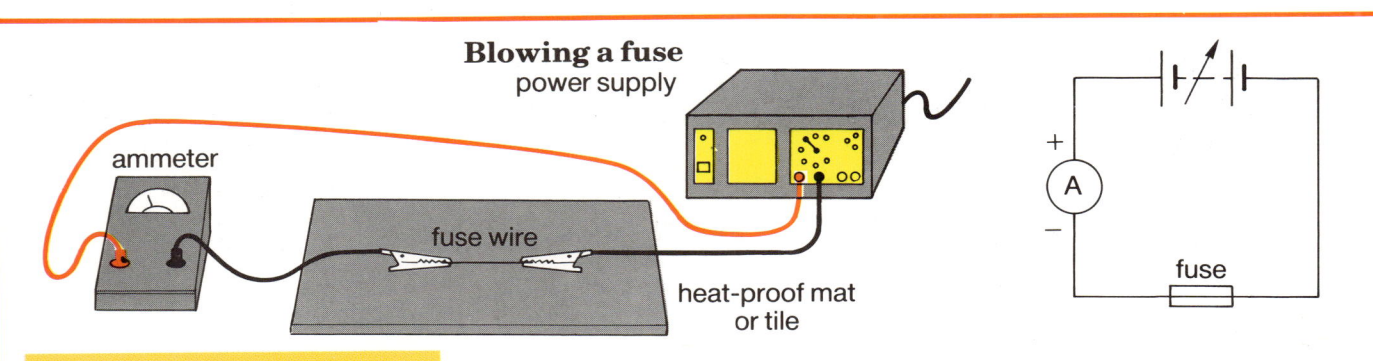

Blowing a fuse

power supply

ammeter

fuse wire

heat-proof mat or tile

+ A − fuse

Collect

- power supply
- ammeter
- connecting leads
- crocodile clips
- bulb
- selection of fuse wires
- heat-proof mat

1 Arrange the fuse wires in order of their thickness. You will use them in this order.
2 Set up the equipment as shown, using the thinnest wire first.
3 Increase the voltage supply slowly. Take care, the wires become hot before they melt.
4 Record the ammeter reading when the fuse wire melts.
5 Repeat with each wire in order.

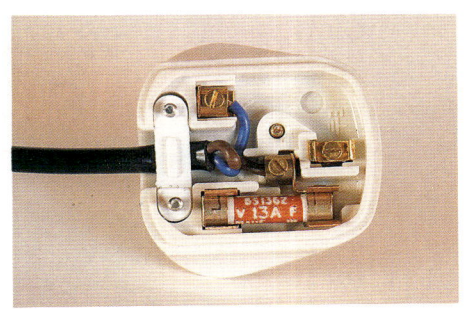

You should already know about the wiring of a plug—if not, look it up in a basic science book like *Understanding Science 1*

Q1 Design a table for your results. What is the pattern? Did all the wires melt?

Q2 Most fuses are either '3 amp' or '13 amp'. This is the current value at which they melt. When you are selecting a fuse for a plug on a particular appliance it must allow enough current to pass to make the appliance work but it should melt if the current becomes higher. Which fuse would you use for the appliances illustrated in Q4 on the previous page?

Joulemeters

The amount of energy used by an appliance depends on its power and on the length of time the appliance is used for. If a hairdryer has a power rating of 1 kilowatt and is used for 15 minutes the energy used is:

$$
\begin{array}{ccccc}
E & = & P & \times & t \\
\text{energy} & & \text{power} & & \text{time} \\
\text{(J)} & & \text{(W)} & & \text{(s)} \\
 & = & 1000\ \text{W} & \times & 900\ \text{s} \\
 & = & 900\ 000\ \text{J} & &
\end{array}
$$

A **joulemeter** can be used to measure directly the energy used by an electrical device.

Collect

- joulemeter
- water heater
- connecting leads
- beaker of water
- power supply

1 Set up this circuit.

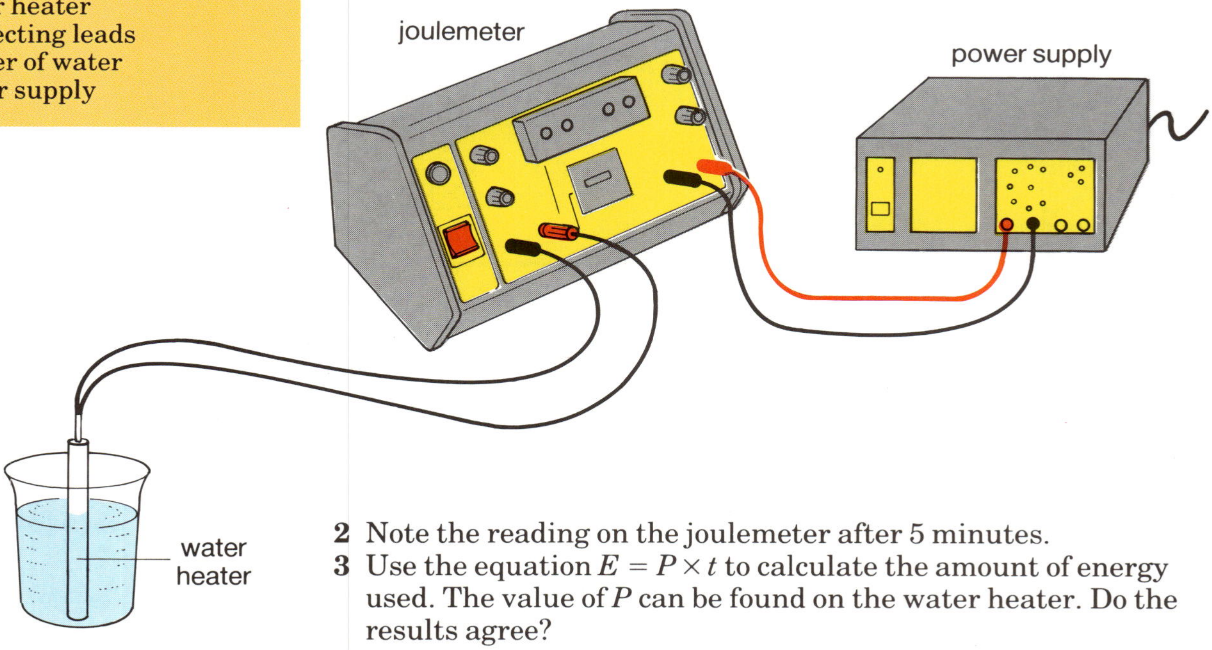

joulemeter

power supply

water heater

2 Note the reading on the joulemeter after 5 minutes.
3 Use the equation $E = P \times t$ to calculate the amount of energy used. The value of P can be found on the water heater. Do the results agree?

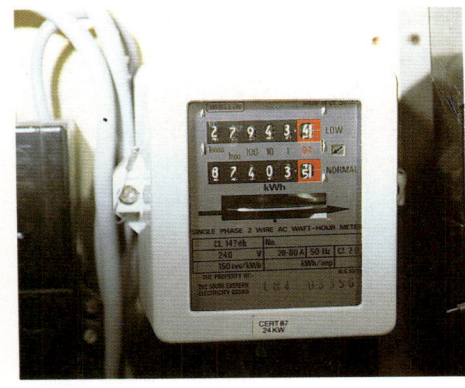

In your home the electricity meter acts as a joulemeter. The meter measures the amount of energy used in **kilowatthours** (kWh). This is a much larger unit than the joule.

energy (kilowatthours) = power (kilowatts) × time (hours)

Kilowatthours are also used as units of electricity on the electricity bill.

Bring this account with you or, if writing or telephoning, quote:-

ELECTRICITY ACCOUNT NUMBER AND DATE (TAX POINT)

510 615 0270 24   18 DEC 1991

Description	Meter Readings		Units Supplied	Pence Per Unit	£
	Present	Previous			
DOMESTIC UNITS TO 17 DEC	12212	11259		7.35	70.05
DOMESTIC QUARTERLY CHARGE					10.84
VAT 0.0% ON DOMESTIC USE OF			£80.89		0.00

FACTSHEETS COVERING WAYS TO PAY, HELPING PENSIONERS & THE DISABLED AND HANDLING COMPLAINTS ARE AVAILABLE AT ALL SEEBOARD SHOPS AND OFFICES.

Q1 Calculate the total number of units of electricity used by the consumer who received this bill.

 Project

It is your task to advise a family on how to reduce their electricity bills. You need to find out how much energy is used by different appliances.

1 Make a list of all the mains electrical appliances your family uses at home.
2 Estimate how many hours each of them is used for in one week.
3 Find out the power rating of each of the devices (make sure they are switched off first).
4 Choose the five appliances which are used for the longest time plus the five which have the highest power rating. Work out how much energy these use in one week altogether.

What is your advice?

Making electricity work

Motors

All of these machines are driven by electric motors. An electric motor is made up of a coil of wire wound round an iron **core**, mounted on a rotating axle and positioned in the field of a **permanent magnet**. When an electric current passes through the wire coil, it produces a magnetic field and the core is magnetised, becoming a temporary **electromagnet**. There are now two magnetic fields, one from the permanent magnet and one from the electromagnet. The way these two fields react with each other produces a force which makes the coil spin.

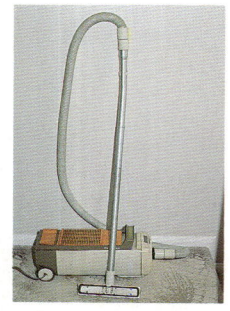

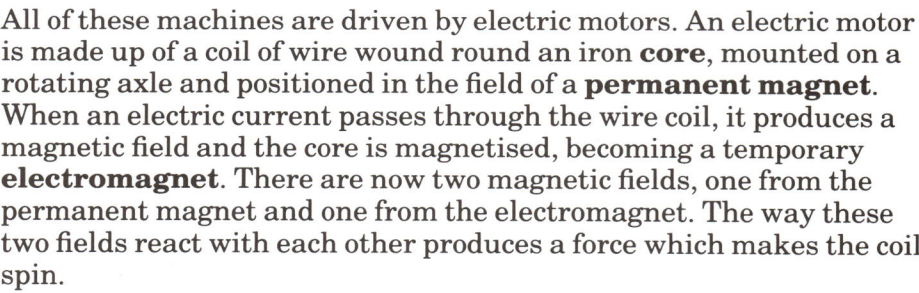

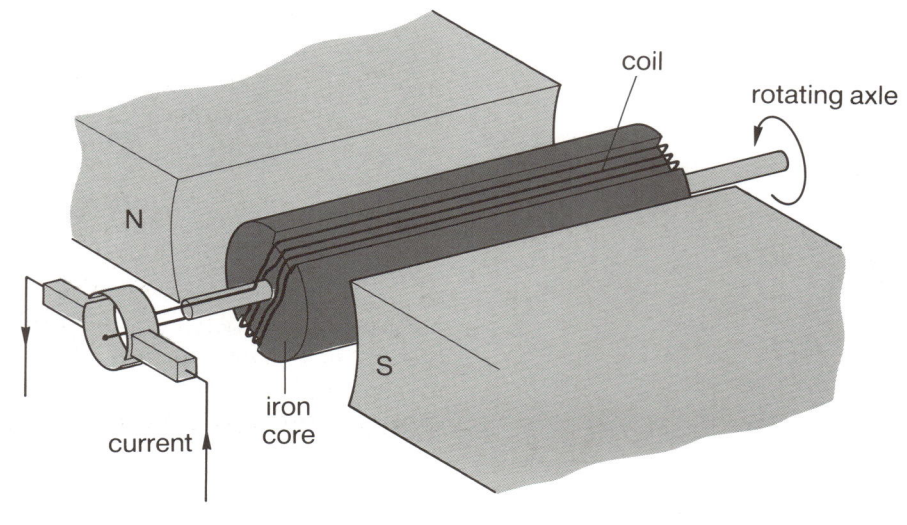

coil

rotating axle

N

S

iron core

current

Make your own motor

Use the pieces in the kit to make the motor shown here.

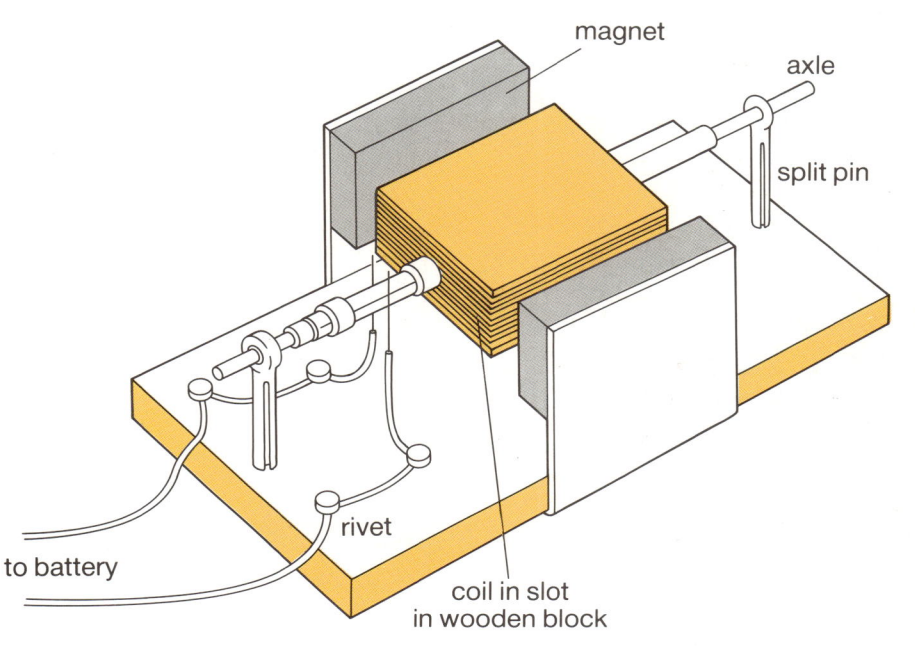

Q1 How could you increase the speed at which the axle rotates?

Q2 What would you have to change to make the axle spin the opposite way?

Generators

If you put electricity into a motor, it spins. If you mechanically rotate the coil or the magnet of a motor it *produces* electricity. When this happens the motor is known as a **generator**.

You are probably familiar with a **dynamo**. When you expend mechanical energy by pushing the pedals of your bicycle, the dynamo axle rotates and a simple generator converts this mechanical energy to electrical energy to light your bicycle lamp.

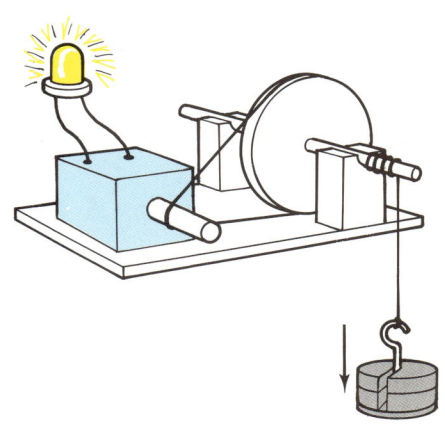

A simple generator

Large-scale generators in power stations produce the electricity we use in our homes.

Most generators produce an **alternating current** (a.c.). This means that the flow of charge periodically reverses its direction in the circuit.

In a power station, the mechanical energy comes from a steam turbine. The steam is produced by burning fuel to heat water

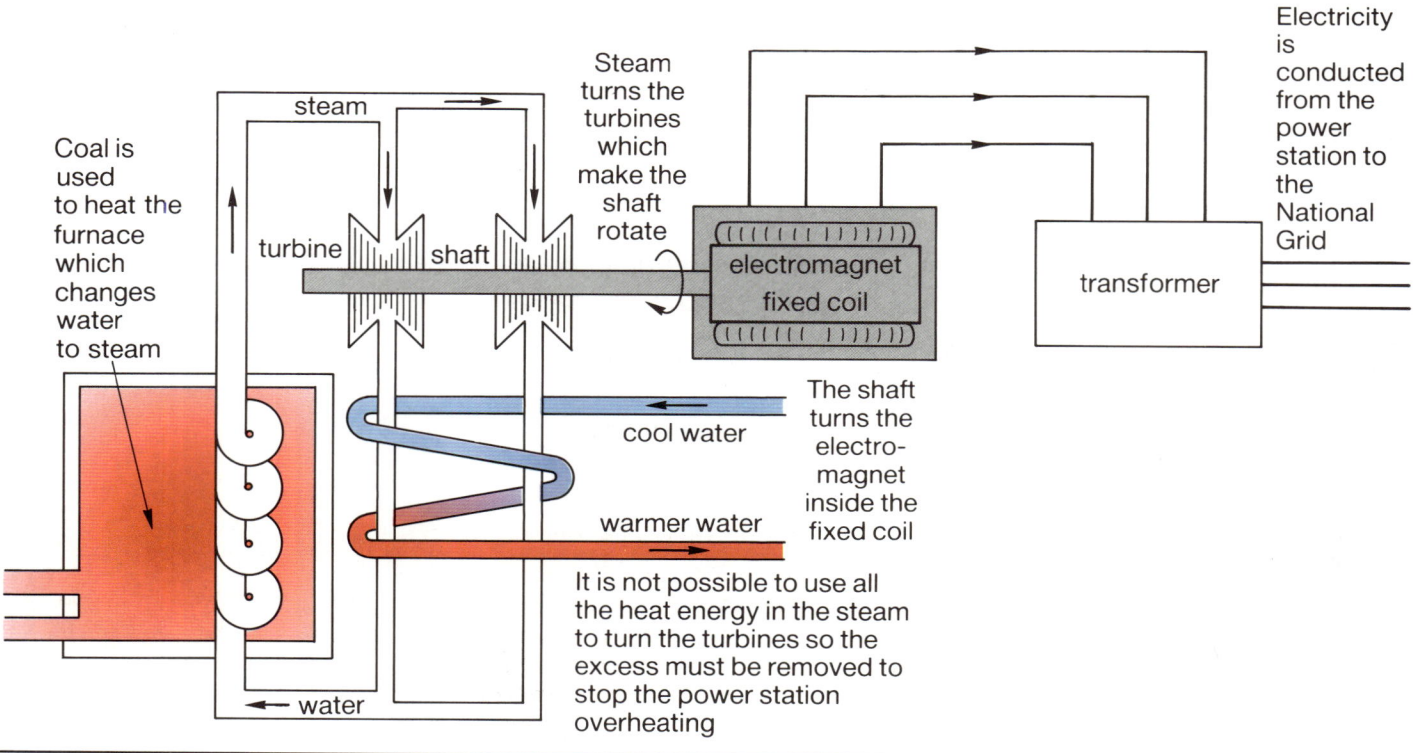

Coal is used to heat the furnace which changes water to steam

steam

turbine   shaft

Steam turns the turbines which make the shaft rotate

electromagnet fixed coil

The shaft turns the electro-magnet inside the fixed coil

cool water

warmer water

It is not possible to use all the heat energy in the steam to turn the turbines so the excess must be removed to stop the power station overheating

water

transformer

Electricity is conducted from the power station to the National Grid

The following diagrams help to explain why a.c. is produced. When the *green* side of the coil is moving *up* through the field (1) the electric current in the circuit flows in one direction. After the coil passes the vertical position (2) the *green* side begins to move *down* through the magnetic field (3) and the current in the circuit is reversed. In the vertical position there is momentarily no current in the coil.

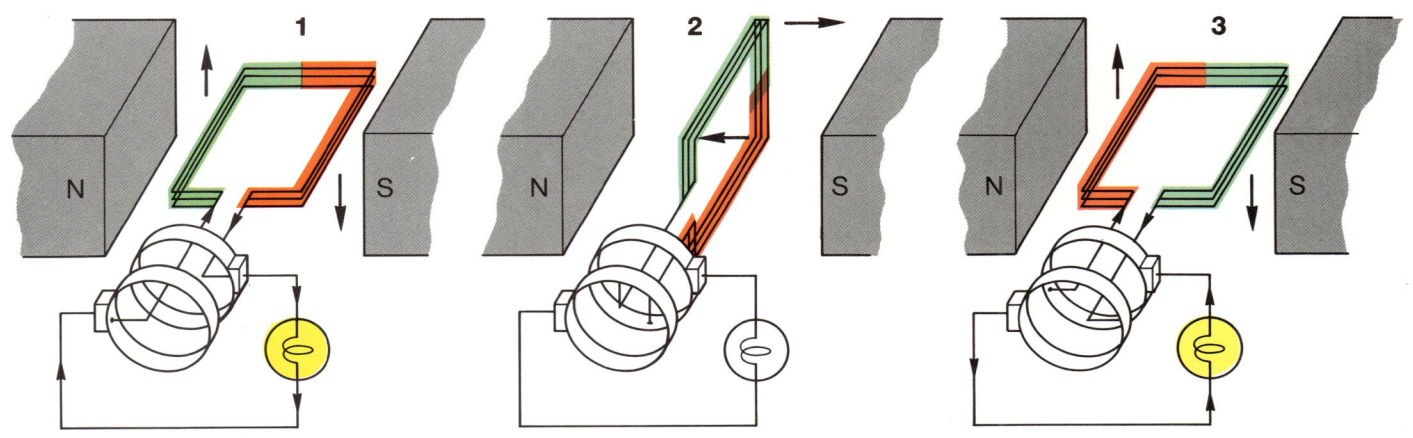

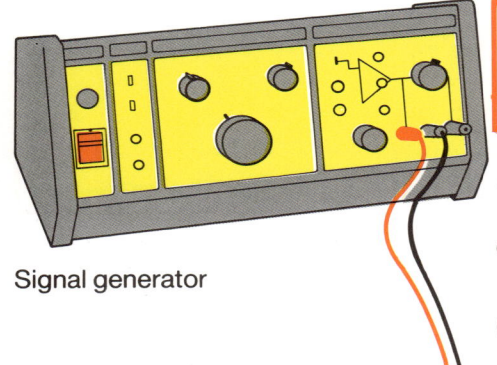

Signal generator

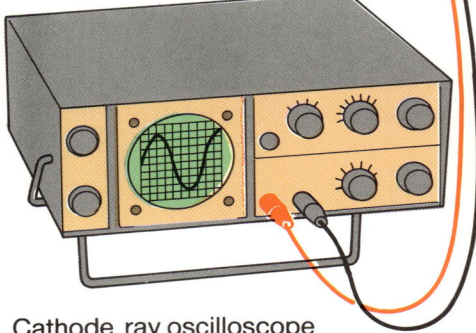

Cathode ray oscilloscope

Your teacher will demonstrate an alternating current using a signal generator and a cathode ray oscilloscope.

Current which does not alternate is called **direct current** (d.c.).

Q1 If the signal generator is replaced with a battery what type of current results?

Q2 If mains electricity is used instead of the signal generator what type of current results?

Q3 What is the frequency of mains electricity?

✍️ *Take notes*

Think about and write down:

- the difference between permanent magnets and electromagnets
- the difference between motors and generators
- the difference between a.c. and d.c.

📚 *Project*

Find out where your nearest power station is. What fuel does it use as its source of energy?

Transformers

Power stations generate electricity at a very high voltage, maybe 10 000 V. The voltage that reaches our home or school is only 240 V. A **transformer** is used to reduce the voltage.

Remember that 'voltage' is a measure of the potential difference (p.d.) in a circuit.

Collect

- transformer kit

Make your own transformer

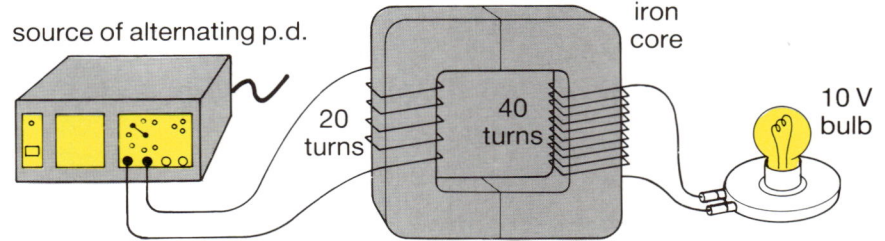

1 Arrange your apparatus as shown.
2 Set the power supply to a p.d. of 5 V and observe the bulb. Gradually increase the p.d. to 10 V and note what happens to the bulb.
3 Decrease the p.d. to 2 V and observe the bulb. Note what happens.
4 Switch off.
5 Now remove the wire coils and replace with 40 turns on the p.d.-source side and 20 turns on the bulb side. Switch on again and note what happens to the bulb.
6 Gradually increase the p.d. to 10 V and note what happens to the bulb.

Q1 Write a report of your experiment.

Q2 Which arrangement of the turns around the core made the bulb shine brightest?

You have made a transformer which can 'step up' or 'step down' the p.d. (voltage) of a power supply.

If the number of turns on the output side (N_2) is greater than the number of turns on the input side (N_1) then the voltage output to the light bulb (V_2) is greater than the voltage input (V_1).

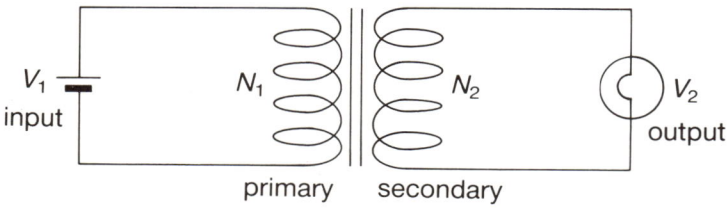

Example a If $N_2 = 40$ turns, $N_1 = 20$ turns, and $V_1 = 5$ V:

$$\frac{V_2}{V_1} = \frac{N_2}{N_1}$$

$$\frac{V_2}{5} = \frac{40}{20} \qquad V_2 = 10 \text{ V}$$

This is a **step-up** transformer.
The same equation is used if N_2 is less than N_1.

Example b If $N_2 = 20$, $N_1 = 40$ and $V_1 = 10$ V:

$$\frac{V_2}{V_1} = \frac{N_2}{N_1}$$

$$\frac{V_2}{10} = \frac{20}{40} \qquad V_2 = 5 \text{ V}$$

This is called a **step-down** transformer.

N_2/N_1 is called the **turns ratio** of a transformer.

As well as being used in the mains electricity supply system, there are smaller transformers in many electrical devices which change the mains supply voltage to the value required by the device.

Some low voltage devices require the use of a separate plug-in transformer

What happens to the current when the voltage is stepped up or stepped down? For the purposes of simple calculations we assume that the power output from the transformer is equal to the power input. In practice there is some energy loss due to heating in the transformer, but this can be reduced to make a transformer more efficient.

$$\text{power } P = I \times V$$

P remains constant, so if V increases, I must decrease.

$$I_1 \times V_1 = I_2 \times V_2$$

$I_1 \rightarrow$ $I_2 \rightarrow$

V_1 V_2

Q1 If $I_1 = 1$ A, what is I_2 in example **a** above?

Q2 If $I_1 = 10$ A, what is I_2 in example **b** above?

✎ Take notes

Discuss with your partner and draw a diagram to show how a step-up and a step-down transformer work. Add brief notes around your diagram to explain what happens.

CHECKPOINT

Power lines

Electricity is distributed around the country in cables supported by pylons. The power stations produce electricity at 10 000 V and 1000 A. If this current was passed along the cables a great deal of energy would be lost as heat. Can you think of a way to reduce the current in the cables as it leaves the power station so that energy loss as heat is reduced?

The power loss can be calculated by

$$P = I^2 \times R$$

where R is the resistance of the cable. We can reduce power losses by keeping I and R low.

To reduce the resistance of the cable we would need to reduce its length, increase its cross-sectional area or make it from a material which is a better conductor. We cannot reduce the length for obvious reasons; to increase the cross-sectional area would be very expensive and the cables would be heavy to support; and the best conducting materials are prohibitively expensive.

It makes sense, therefore, to reduce power loss by reducing the *current*, using a step-up transformer.

Demonstration of energy loss in a power line

Collect
- power line kit
- power supplies
- transformers

1 a Set up the apparatus and set the power supply to 2 V.
b Observe the bulb.
c Gradually increase the power to 5 V and observe the bulb.

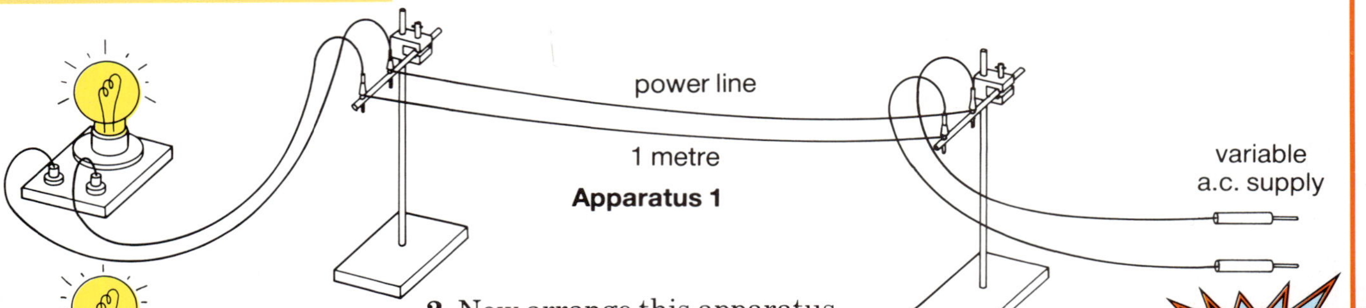

power line

1 metre

Apparatus 1

variable a.c. supply

2 Now arrange this apparatus.

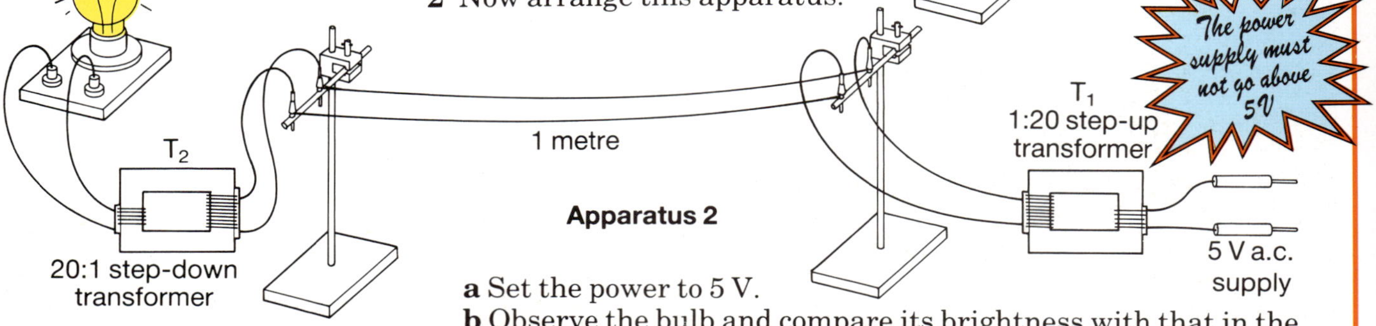

T_2

T_1
1:20 step-up transformer

The power supply must not go above 5 V

20:1 step-down transformer

1 metre

Apparatus 2

5 V a.c. supply

a Set the power to 5 V.
b Observe the bulb and compare its brightness with that in the first arrangement.

Transformers at a power station increase the voltage to 250 000 V or greater for transmission over long distances. At the destination transformers at a substation will decrease the voltage as required— to 240 V for houses, schools, shops, offices etc.

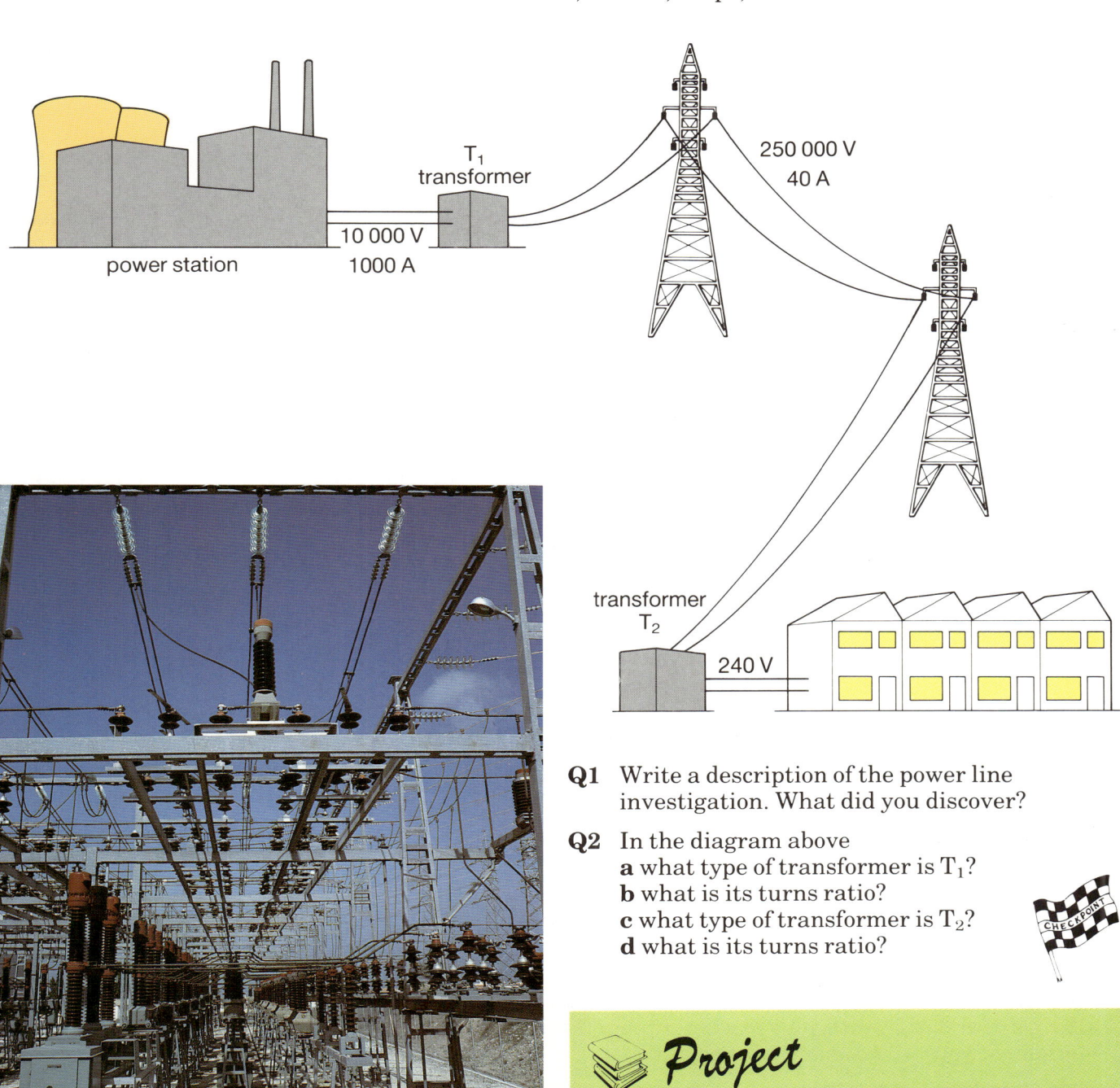

Rows of transformers at a substation

Q1 Write a description of the power line investigation. What did you discover?

Q2 In the diagram above
  a what type of transformer is T_1?
  b what is its turns ratio?
  c what type of transformer is T_2?
  d what is its turns ratio?

Project

Find out about the National Grid system; or, if you live in Scotland or Northern Ireland, about your area's generation and distribution system.

Portable energy

Batteries are a very useful way of storing electrical energy. They can be small, light and thus easy to carry, and most have sufficient potential to provide electricity for a reasonable period of time.

A battery consists of a group of 'electrolytic' **cells**. A 'dry' cell consists of two **electrodes** with a paste of chemicals, called the **electrolyte**, between them.

The electrodes and the electrolyte react chemically with one another to produce **ions**. There is a build-up of negative charge on one electrode, the **cathode**, and a build-up of positive charge on the other, the **anode**. The charge is stored until the battery is connected into a conducting circuit. Then a flow of electrons is set up from cathode to anode, the electrical energy coming from the stored chemical energy of the battery.

The battery in a car provides the electrical energy to produce the ignition spark (as well as for lights, windscreen wipers etc.). The car runs by vaporised petrol activating pistons which rotate the axles.

There is presently interest in developing a battery that could drive the motor of a car, as in a milkfloat. Pollution from exhaust gases would cease and the motor noise would be reduced. However, at the present time, a battery powerful enough to run a car would be too large for practical use and would also need recharging approximately every 40 miles. But this is not *your* problem!

Problems with batteries

Work in a group and design experiments to investigate one or both of these problems.

1 Does temperature affect the length of time that a torch battery lasts?
2 How can a fruit or vegetable be used to make a battery which can power a small device such as a clock?
Which fruit or vegetable provides the most power?

You will need to consider:

- how you will measure the life of a battery
- which variables you will need to keep the same
- safety factors
- how you will present your results.

Discuss your design with your teacher, then collect the equipment you will need and investigate!

Deliver a presentation to the class describing why you designed your equipment in the way you did, what happened and the conclusions you have made.

Are there any practical applications of your findings?

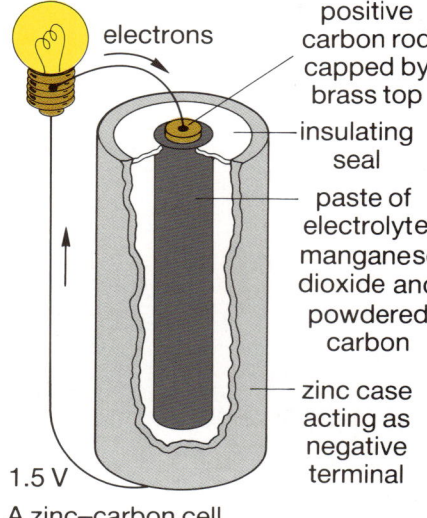

A zinc–carbon cell

- electrons
- positive carbon rod capped by brass top
- insulating seal
- paste of electrolyte: manganese dioxide and powdered carbon
- zinc case acting as negative terminal

1.5 V

Batteries that are cheapest to buy have zinc–carbon cells

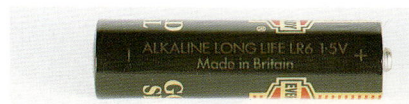

'Long-life' batteries have alkaline–manganese cells

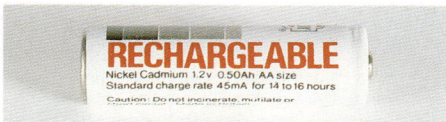

Small rechargeable batteries have nickel–cadmium cells. They need to be recharged, by mains electricity sending electrons in the opposite direction

Large rechargeable batteries have lead and lead dioxide electrodes and a liquid electrolyte (sulphuric acid). An example is the car battery

Talkabout

Communication

- Discuss the advantages and disadvantages of each form of communication.

- Why is the fax machine so popular with businesses?

1 An electric iron is fitted with an earth and a plug containing a fuse as shown below.

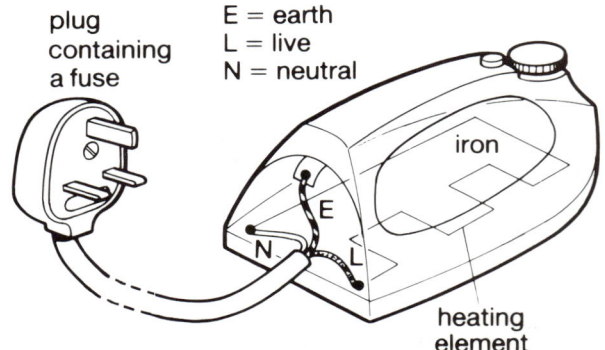

plug containing a fuse

E = earth
L = live
N = neutral

iron

heating element

a Describe briefly how a fuse works. [2]

b Give **two** reasons why a fuse in a plug is important to the safe use of a mains-operated appliance. [2]

c Why should the fuse in the plug be connected to the live wire? [1]

d Explain how earthing the iron protects the user from receiving an electric shock. [3]

e The iron is rated at 1500 W, 240 V.
  i Calculate the size of the current through the iron. [3]
  ii You are provided with the following fuses:

    1A   3A   5A   13A

  Which one would be safe to use in the plug? [1]

NEA 1989 Physics

2 Decorative tree lights can be arranged in two ways as shown in the diagrams below.

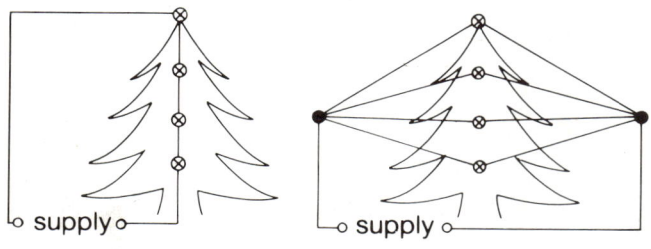

Circuit A: bulbs in series   Circuit B: bulbs in parallel

If one bulb blows in each of the circuits, in which circuit will the remaining bulbs stay alight? [1]

b A diagram of a typical light bulb is shown below.

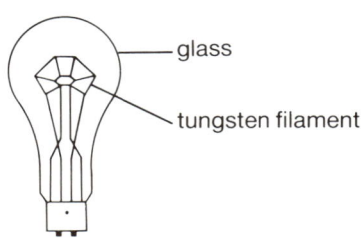

glass

tungsten filament

  i In the light bulb electrical energy is converted to ...... and .... [2]
  ii Suggest one property which makes tungsten a suitable metal to use for the filament. [1]

c The diagram below shows a three-pin plug.

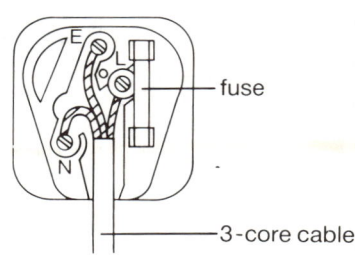

fuse

3-core cable

  i What important safety feature, other than the top, is missing? [1]
  ii The fuse used in a plug should be suitable for the appliance connected to it. Using the following relationship:

    power measured in watts = current in amps × voltage in volts

  complete the table. Choose the most suitable fuse from 3 A, 5 A and 13 A.

Appliance	kettle	video recorder
Power rating in watts	3000	50
Voltage supplied in volts	250	250
Current rating of the most suitable fuse in amps	[1]	[1]

MEG 1988 Specimen Paper Science Syllabus A

5

Energy and chemistry

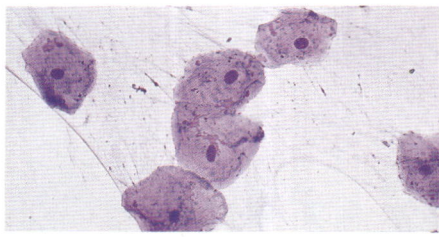
Energy needed by living things is released in cells

Solid fuel cookers burn coal to provide useful heat

An energy circus
With your partner, visit each of the stations and do the experiment described on the card. The card gives instructions for each experiment and the information and observations that you need to record. Make sure you have written your answers before you move on to the next station. Ask your teacher if you are unsure about anything.

Energy released!

Each of the photographs shows an example of energy being released in a chemical reaction.

A glow-worm undergoing chemiluminescence

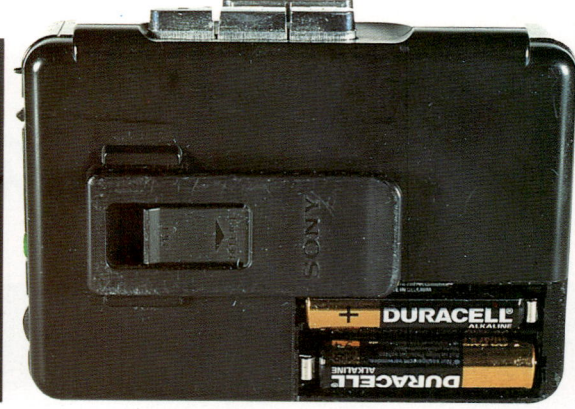
The chemical reaction in a battery produces electrical energy

An oil well on fire releases huge amounts of heat

 *Take notes*

Discuss with a partner and write down a summary of your observations. This should include:

- a list of those reactions which gave out energy and the type of energy released
- a list of those reactions which absorbed energy and the type of energy absorbed.

A rechargeable battery

When a battery is charged up, a chemical reaction takes place during which electrical energy is stored. When the bulb is connected the electrical energy is released and the chemical change reversed.

Collect

- beaker
- 2 lead plates
- dilute sulphuric acid
- power pack
- leads and crocodile clips
- switch
- bulb
- timer
- safety glasses
- graph paper

Charging up

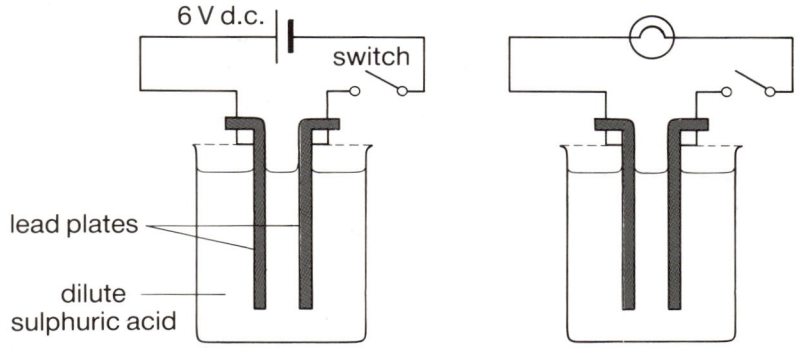

1 Set up the circuit on the left.
2 Switch on the power and charge up your battery for about thirty seconds.
3 Disconnect the power pack and connect the bulb.
4 Time how long the bulb stays lit.

You could investigate the effect of
- charging the battery for different times to see how this affects the time the bulb stays lit
- a different distance between the lead plates
- different depths of acid.

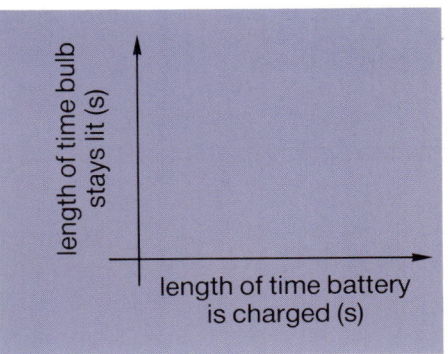

Q1 Write a report on your investigation. Present your results in the form of a table and draw a graph of them.

[IT] Use a data-handling program to plot the graph.

Focusing on heat changes

Each of these reactions releases heat to the surroundings as new chemical compounds are formed.

dilute sulphuric acid + dilute sodium hydroxide solution → sodium sulphate solution + water + HEAT

$H_2SO_4(aq) + 2NaOH(aq) \rightarrow Na_2SO_4(aq) + 2H_2O(l)$

hydrogen gas + oxygen gas → water + HEAT

$2H_2(g) + O_2(g) \rightarrow 2H_2O(l)$

aluminium + iron oxide → iron + aluminium oxide + HEAT

$2Al(s) + Fe_2O_3(s) \rightarrow 2Fe(s) + Al_2O_3(s)$

The reaction of aluminium and iron oxide

Reactions which release heat are called **exothermic** reactions.

Some reactions absorb heat from the surroundings as new chemical compounds are formed.

HEAT
released

REACTANTS ⟶ PRODUCTS

HEAT
released

HEAT
absorbed

REACTANTS ⟶ PRODUCTS

HEAT
absorbed

Reactions which absorb heat are called **endothermic** reactions.

We can show the energy changes in energy level diagrams.

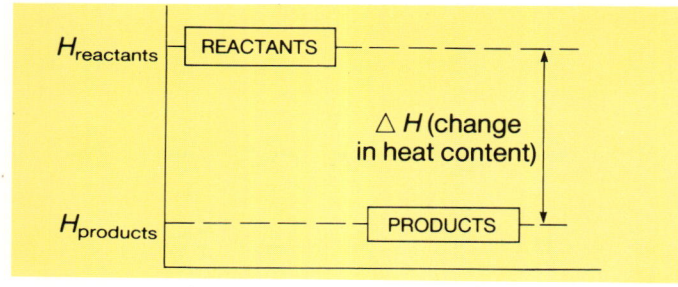

$H_{reactants}$ — REACTANTS

$\triangle H$ (change in heat content)

$H_{products}$ — PRODUCTS

a   Exothermic reaction

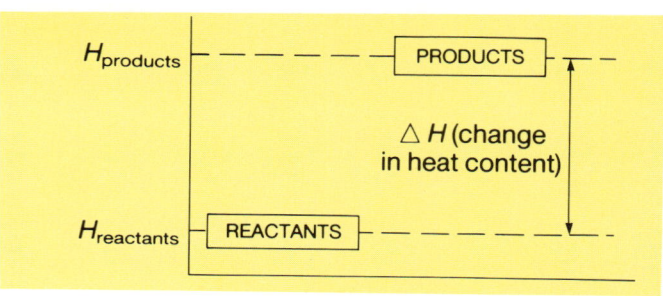

$H_{products}$ — PRODUCTS

$\triangle H$ (change in heat content)

$H_{reactants}$ — REACTANTS

b   Endothermic reaction

The amount of stored energy (the heat content) is given the symbol H. The energy change (heat change) is given a special symbol, ΔH (delta H).

$$\begin{aligned} \text{heat change} &= \text{heat content} - \text{heat content} \\ \text{of the reaction} & \quad \text{of products} \quad \text{of reactants} \\ \Delta H &= H_{\text{products}} - H_{\text{reactants}} \end{aligned}$$

If $H_{\text{reactants}}$ is larger than H_{products} then ΔH is negative and the reaction is exothermic.

If $H_{\text{reactants}}$ is smaller than H_{products} then ΔH is positive and the reaction is endothermic.

Certain types of chemical reactions can provide us with **usable** energy. This can be used directly or converted to a more convenient form. Coal, oil and gas are fossil fuels. Fossil fuels are formed in the Earth's crust from the remains of plants or microscopic animals. Over periods of hundreds of thousands of years, pressure built up in the Earth's crust and transformed the remains into coal, oil and gas. **Burning** fossil fuels provides us with much of the energy that we use in the home, in industry, at work or for recreation.

In order to provide useful energy, a chemical reaction should:

- provide large amounts of energy
- release the energy quickly—but not too quickly—we must be able to control the rate
- be cheap enough to use on a large scale—it should not use large amounts of scarce or expensive resources
- not cause pollution.

Q1 Collect a large, complete copy of the diagram showing different types of energy. Think of applications for each of the different types marked A–E and write them on the diagram. Stick the diagram in your book.

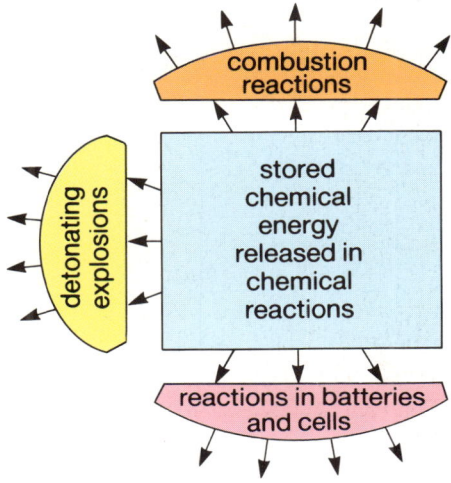

Q2 Draw one energy level diagram for an exothermic reaction and one for an endothermic reaction.

Q3 Look for information on the topic, 'How we use energy in today's world.' You could use information from other books, photographs from magazines, newspaper articles or some other source.

Q4 Make a poster to show the importance of chemical reactions in meeting our energy needs.

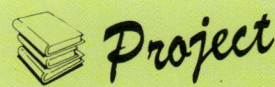

Project

Find out how chemical explosives are used in industry.

What happens when fuels burn?

Collect

- bunsen burner
- two heat-proof mats
- test tube holder
- three large test tubes
- cobalt chloride paper
- candle
- firelighter
- wooden splint
- paraffin
- ethanol
- metal tray
- watch glass or tin lid
- glass wool
- safety glasses

Be careful:

- burn only one fuel at a time
- follow the instructions for liquid fuels
- keep anything which might catch fire away from your experiment
- make sure all flames are properly put out
- keep your glasses on at all times.

1 Light the bunsen burner and then turn the gas on fully.
2 Adjust the air hole to get different flames. Try altering the amount of gas as well.
3 Observe each flame and describe it. Test each flame by holding a clean test tube containing cold water just above the top of the flame. If you see any 'misting up' or droplets of clear liquid try touching it with some anhydrous cobalt chloride paper. Describe anything else which appears on the test tube.
4 Do the same for a burning wooden splint, firelighter and candle.
5 Burn the liquid fuels and test them in the same way.

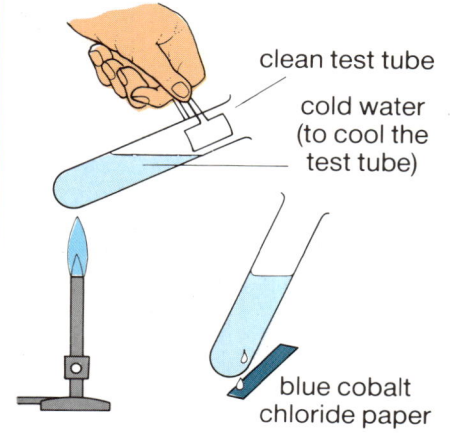

clean test tube

cold water (to cool the test tube)

blue cobalt chloride paper

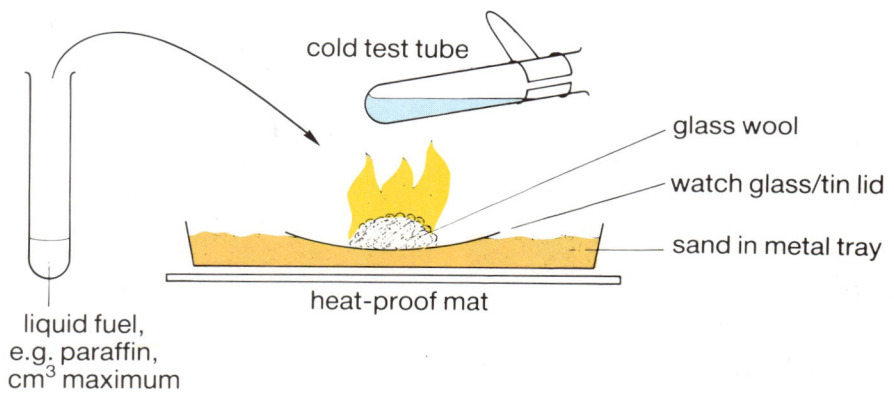

cold test tube

glass wool

watch glass/tin lid

sand in metal tray

heat-proof mat

liquid fuel, e.g. paraffin, 1 cm³ maximum

- Clean the watch glass/tin lid after each fuel (**care**: hot apparatus)
- In an emergency put the flame out by covering the apparatus with a heat-proof mat

Q1 Draw a table to record your observations.

Q2 Which of your fuels are *clean* and which are *dirty*? Can you offer any explanations?

Q3 Solid and liquid products are fairly easy to notice. Do you think there are other kinds of products which you haven't observed? If you think there are, suggest how these could be detected—and maybe identified.

Q4 Can you see any pattern in your observations?

What is produced when methane burns?

Your teacher will demonstrate this experiment.

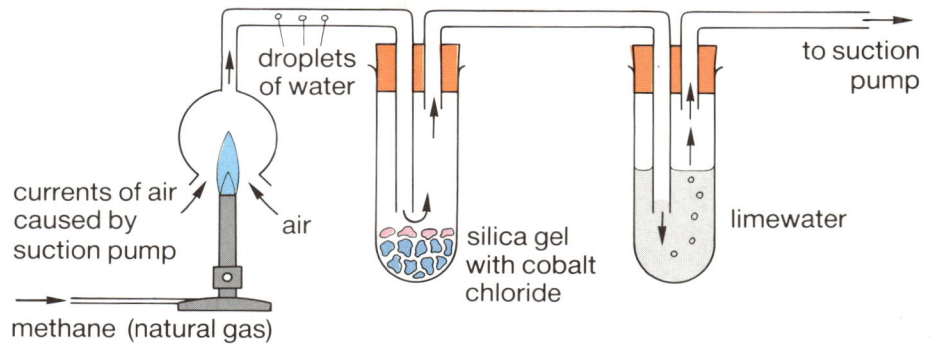

Q5 Why are cobalt chloride and limewater used in the experiment?

Q6 What is produced when methane burns?

Q7 Your teacher may pass air through the experiment without burning a fuel. Describe what happens. How does this affect your conclusions about the products of combustion?

Making a gas

Cracking oil

Crude oil is an important source of fuels (as well as raw material for manufacturing). Crude oil itself is very difficult to burn without causing a great deal of pollution. Crude oil can be broken down to smaller molecules. This is called **cracking**. The products are often gases, which mix more easily with air and are easier to burn.

Collect

- large test tube
- 2 small test tubes
- pumice or alumina powder or broken pot
- clamp and stand
- heat-proof mat
- bunsen burner
- delivery tube and bung
- bunsen valve
- trough of water
- paraffin oil
- rockwool
- safety glasses

CARE NEEDED! Fire risk!

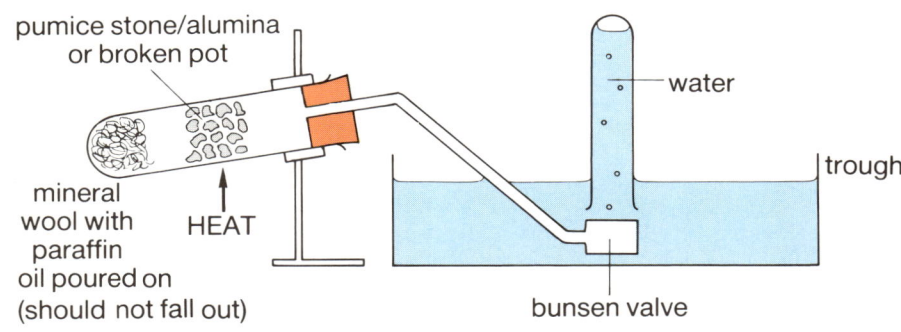

1 Set up the apparatus.
2 Ask your teacher to check your apparatus before you start.
3 Heat the pumice stone (broken pot/alumina) strongly. Occasionally heat the crude oil for a second or two only.
4 Collect two tubes of gas.
5 Try to light the gas and observe what happens.
6 Try to find out if water and carbon dioxide are produced when the gas is burned.

Key

carbon C

oxygen O

○ hydrogen H

These colours are used by convention and will be used throughout this book

Q1 Report your experiment fully.

Q2 What do you think the products of combustion are?

Q3 Why is gaseous fuel easier to burn?

Q4 Suppose you had to decide if your product would make a good fuel. What would you have to consider before deciding whether it is worthwhile producing it on a large scale?

All the commonly used fuels—coal, oil and gas—contain chemicals called **hydrocarbons**. All hydrocarbons are made up from carbon and hydrogen. The first four members of the hydrocarbon series are:

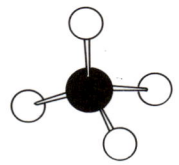

methane CH₄

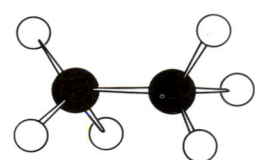

ethane C₂H₆

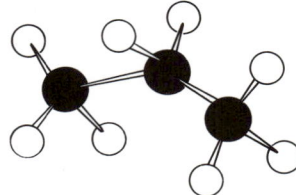

propane C₃H₈

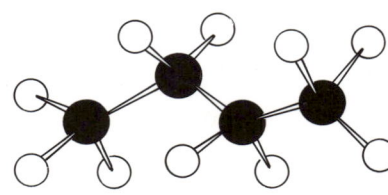

butane C₄H₁₀

Methane is the main chemical compound in **natural gas**. Ethane is another gaseous fuel. Propane and butane can be liquified and stored in pressurised metal tanks or bottles. Butane is widely used, bottled, for camping stoves

When methane burns in plenty of oxygen the oxygen combines with the carbon and hydrogen to form **carbon dioxide** and **water**. These are the products of combustion.

methane	+	oxygen	→	carbon dioxide	+	water
$CH_4(g)$	+	$2O_2(g)$	→	$CO_2(g)$	+	$2H_2O(g)$

When propane burns, the products of combustion are the same. However, more oxygen is needed because there are more carbon and hydrogen atoms in a molecule of propane than there are in a molecule of methane.

propane	+	oxygen	→	carbon dioxide	+	water
$C_3H_8(g)$	+	$5O_2(g)$	→	$3CO_2(g)$	+	$4H_2O(g)$

The combustion of hydrocarbons is an example of an **oxidation** reaction—carbon dioxide and hydrogen oxide (water) are formed as the hydrocarbon combines with oxygen.

Complete combustion of hydrocarbons leads to carbon dioxide and water formation only and occurs when there is an adequate supply of oxygen. **Incomplete combustion** occurs when the air supply is cut down and there is not enough oxygen to combine with all the carbon and hydrogen atoms. Some carbon monoxide ($C\equiv O$) will be formed as well as carbon dioxide and water. Some carbon may also be formed. It is dangerous to burn fuels without a good supply of air because carbon monoxide is poisonous. If breathed in in quantity it prevents the blood from carrying oxygen to all the cells in your body—including your brain cells—and causes death. The carbon forms soot and smoke which as well as being dirty is harmful to our lungs and can damage our health.

carbon monoxide

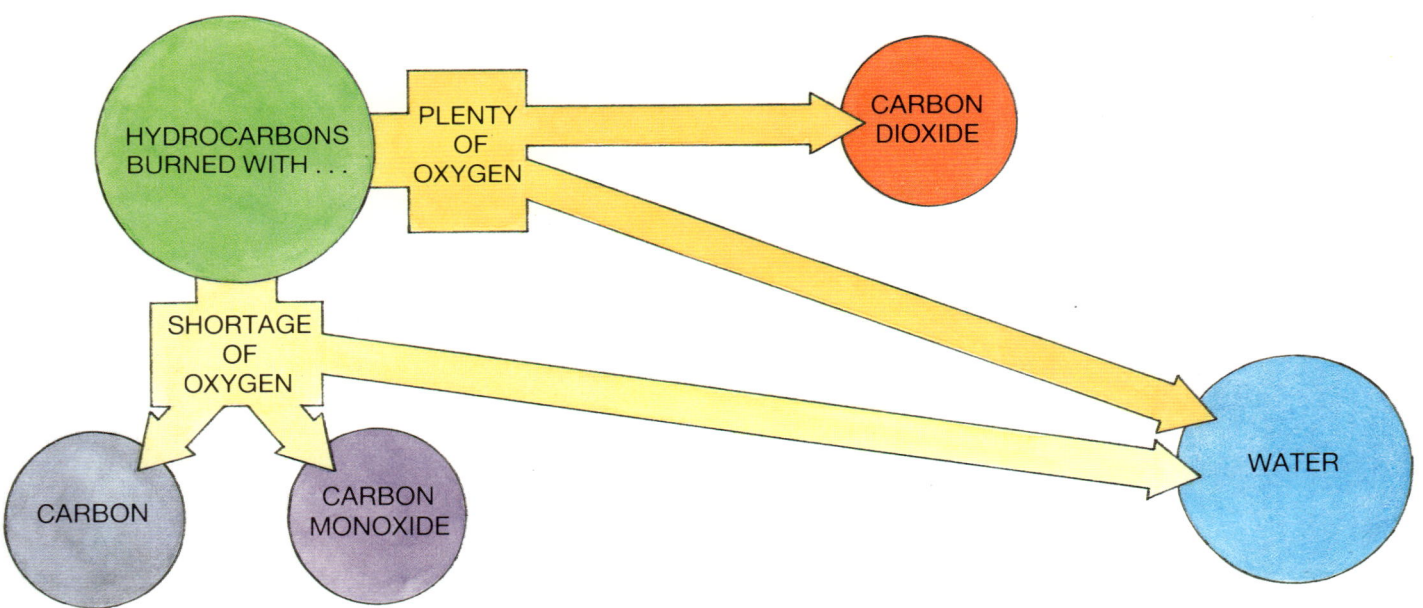

Fuels like coal and oil contain other elements besides carbon and hydrogen. Sulphur is often present and combines with oxygen, when the fuel is burned, to form sulphur dioxide gas. Sulphur dioxide dissolves very easily in water to form an acidic solution—sulphur dioxide is the pollutant that forms acid rain.

Discuss with your partner and write down:
- what is meant by a hydrocarbon
- the differences between complete and incomplete combustion
- why incomplete combustion is dangerous
- why combustion is an oxidaton reaction.

Project

Find out the names and locations of the North Sea oil and gas fields. Draw a map showing their positions.

Choosing a fuel

Turning on the heat

How do we decide on the best fuel for our purpose? Fuels can have many different properties—we shall be looking at some later—one of the most important is how much heat is available from one kind of fuel compared to another?

Collect

- thermometer
- tin can
- tripod
- stopwatch
- liquid fuel burners
- measuring cylinder
- two or three heat-proof mats to form a draught shield
- balance
- safety glasses

Compare the heat output of paraffin and ethanol
Liquids can be burned easily in this apparatus.

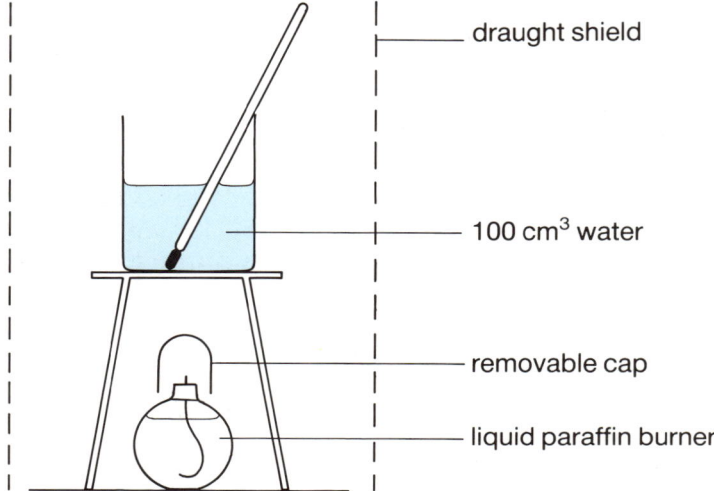

- draught shield
- 100 cm³ water
- removable cap
- liquid paraffin burner

The cap is useful for putting the flame out and stopping the fuel from evaporating when you are not using it.

You must decide what evidence you will use to say which liquid fuel has the greatest heat output.

1 Set up the apparatus.
2 Weigh the burner and note the mass.
3 Take the temperature of the water.
4 Light the burner and heat the water.
  Try to make sure that your results are reliable. (**Hint**: do you think that you will get the same result if you carry out the experiment again?)

Q1 Report on what you have found out.
  a Make a note of any changes you made to the results you collected.
  b Write down any ideas you have about simple improvements to your experiment.

(**Hints:**
- did the fuel convert all its stored energy into heat energy?
- was all the heat of the flame trapped in the way you intended?)

How much energy?

The amount of energy released in a chemical reaction is measured in **joules**. The joule is the scientific unit of energy just like a gramme is the scientific unit of mass and metre is the unit of length.

The joule (J) has a very exact definition. 4.2 J of energy will raise the temperature of 1 g of water by 1 °C. So 4.2 kilojoules (kJ) are needed to warm up 1000 g (1 kg) of water by 1°C.

When you have done your experiment to compare the heat output of different fuels you can convert the temperature rise of the water into joules of energy.

$$
\begin{array}{lll}
\text{energy} & = \text{temperature rise} \times \text{mass of water} \times 4.2 \\
\text{(joules)} & \quad (°C) & \qquad\qquad (g)
\end{array}
$$

What makes a good fuel?

We must decide what we are looking for in a fuel. Obviously it must produce all the heat energy we require. Ideally it should have a very high heat content per kilogram. It should also be

- plentiful and very versatile

- easy to burn

- convenient to store

- easy to transport

- pollution free and non-toxic and should harm the environment as little as possible.

Nobody has yet found a perfect fuel! Wood is the oldest fuel—in many parts of the world it is easy to obtain in fairly large quantities. It burns readily and can provide reasonable heat. Although it is a good fuel for small-scale and medium-scale use it has serious disadvantages for industry and for our environment.

- A lot of wood needs to be burned to get sufficient heat for large-scale applications.

- It produces a lot of ash which can choke fires and slow down burning. The ash needs to be removed.

- It is inconvenient to store large amounts of wood and it is difficult to transport.

- It often produces large amounts of smoke.

Coal is a better fuel, mainly because it has a greater heat content per kilogram. It has several disadvantages though.

- It is not readily available in large quantities. It has to be mined.

- Mining causes environmental problems as well as being a dangerous activity.

- It produces smoke and sulphur dioxide. The latter reacts with water in the atmosphere and produces acid rain which is damaging many of our forests and lakes. Pollution is a problem that has to be dealt with.

- It leaves behind large quantities of ash and embers after it has been burned.

- It is bulky and is difficult to store and transport.

- It is valuable as a source of useful chemicals as well as a source of energy.

- It is a non-renewable resource—it does not provide an unlimited supply. An increased demand on coal supplies in the future could lead to severe shortages.

Liquid and gaseous fuels are easier to burn cleanly and do not leave ash. They also have a high heat content per kilogram. The main disadvantages are problems of transport and storage (gas needs special pressurised tanks) and their higher cost. Natural gas is cheaper than other gaseous fuel but of course is a non-renewable resource.

All hydrocarbon fuels produce carbon dioxide which is considered to be a major cause of **global warming** (the Greenhouse Effect).

Take notes

Discuss with a partner and write down:

- the characteristics of a good fuel
- the advantages and disadvantages of coal, petrol and natural gas.

Which is the best fuel?

Collect

- a resource sheet showing the amount of heat energy available from the different fuels and the cost of each

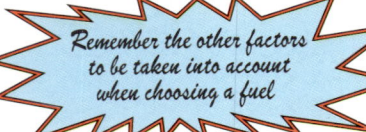
Remember the other factors to be taken into account when choosing a fuel

1 Imagine that you are writing a newspaper article for the *Homebuyers* section of your local newspaper. Advise the reader whether to choose gas-fired central heating or oil-fired central heating.
2 Which would be the best fuel for a new design of energy-efficient car? There are many things you need to consider, for example the car should have good acceleration but be economical with fuel.

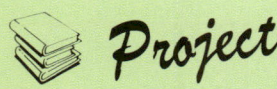
Project

Find out some of the uses of crude oil other than as a fuel.

5.4 *Problem*

Electricity from chemicals

You might have already used electrochemical cells. For revision do the first experiment below.

Collect

- two beakers (100 cm³)
- filter paper
- copper sulphate solution
- other salt solution
- voltmeter (0–5 V)
- copper, zinc, magnesium, iron, lead

1 Use zinc as electrode A.

Find out what happens to the voltage when you use each of the other metals as electrode A.

2 What happens when both electrodes are the same?

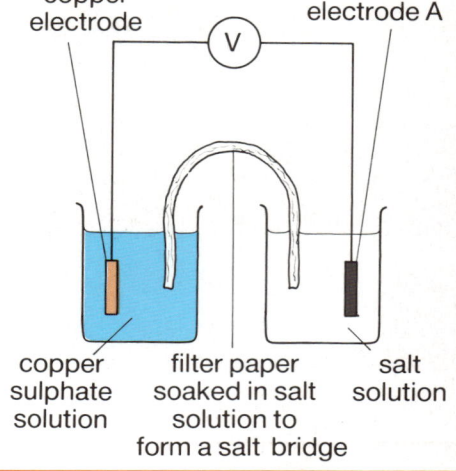

copper electrode

electrode A

copper sulphate solution

filter paper soaked in salt solution to form a salt bridge

salt solution

Q1 Draw your own diagram of the experiment.

Q2 Write your results in a table.

Q3 Explain how your results are related to the reactivity series. (If you have forgotten about the reactivity series look it up in a science book such as *Understanding Science Book 3*.)

Collect

- beakers
- copper electrodes
- glucose solution
- methylene blue
- filter paper
- sensitive voltmeter

An unusual cell

Some years ago, some researchers said that living yeast cells could also provide a potential difference (voltage).

1 Set up the experiment. You need a sensitive voltmeter measuring millivolts.

2 Devise some experiments to show that the potential difference is probably caused by living yeast cells which are respiring (breaking down glucose to release energy).

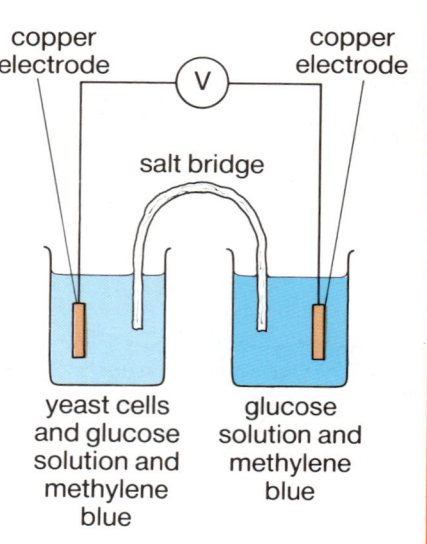

copper electrode

copper electrode

salt bridge

yeast cells and glucose solution and methylene blue

glucose solution and methylene blue

Q1 Choose a suitable title.

Q2 Write a report on your investigation.

Where does it all go?

Energy can be stored and it can be released. Energy can change from one form to another and do work. But energy cannot be destroyed.

If energy cannot be destroyed

- why is there an 'energy crisis'?
- can we run out of energy?
- can we use all forms of energy?
- what happens to all the energy that is pouring onto the Earth from the sun?

1 The major source of the world's energy supply is fossil fuels. The chemical energy that they contain is converted to other more useful forms, e.g. heat and electrical energy. During the last thirty years there has been a considerable increase in demand for electrical energy and now other sources of energy are being investigated as a matter of urgency.

  a Explain what is meant by the term 'fossil fuel'. [1]

  b Name three fossil fuels in use at the present time, each one existing in a different state of matter at room temperature and pressure (a solid, a liquid and a gas). [2]

  c Name the two elements which are present in the highest proportion in fossil fuels. [1]

  Write the symbol equations for the complete combustion of each of these elements in oxygen. [2]

  d Both of the reactions represented in part c are exothermic. Draw an energy level diagram to show the energy changes which take place when one of the elements is burned in oxygen. Indicate clearly on the diagram the value for the heat of combustion of the element by labelling it ΔH. [2]

  e Why are urgent efforts being made to find sources of energy as an alternative to the use of fossil fuels? [1]

  f State one source of energy, other than direct solar energy, which may be used as an alternative to fossil fuels. [1]

  JMB 1983   Chemistry

2 a Describe an experiment which could help you to decide which one of four different fuels gives out the most heat. You may draw diagrams. Say how you would make the tests as fair as possible. [6]

  b List **two** things you would look for when choosing the *best* fuel for heating a home. [2]

  c Name one pollutant formed when fuels are burned. Say what effect it has on the environment. [2]

  MEG 1988   Specimen Extension Paper Science

3 This question is about fuels and energy. Petrol is not the only fuel that can be used in a car engine. Other fuels have been tried, including methanol and ethanol. The table gives some properties of each of these possible fuels. (Energy content simply means how much thermal energy you get from a certain amount of material when it burns.)

Fuel	Boiling point (°C)	Energy content (MJ/kg)
Methanol	64	19.6
Ethanol	80	26.8
Petrol	60	47.9

  a Using the data in the table:
  i give **one** advantage of using petrol rather than methanol or ethanol and explain your answer. [2]
  ii give **one** disadvantage of using ethanol compared with the other two fuels and explain your answer. [2]

  b Instead of petrol engines, some cars are fitted with diesel engines. Diesel oil goes solid at $-10\,°C$. Give **one** reason why using diesel oil could be a disadvantage. [1]

6

Energy and force

Energy changers

Conserve your energy

Energy is needed by all living things. Food and sunlight provide energy to make things grow, to provide warmth and to sustain life. The human population also requires energy to provide artificial light and heat and to run machines. As the population of the world increases and Third World countries develop, more and more energy is required. However, the fossil fuels on which we rely for energy, such as coal, oil and gas, are likely to run out in the not too distant future. One of the most important tasks facing engineers and scientists today is to find new energy sources and to find ways of using those we have more efficiently and sensibly.

The cartoons on the left show two examples of using energy.

In both these cases energy is released as a result of chemical reactions. Kinds of energy include heat, kinetic, potential, sound, heat, light, magnetic, electrical, gravitational and nuclear.

You get energy from the chemical energy in food. A car gets energy from the chemical energy in the petrol

chemical energy (food)
to kinetic energy (movement)

kinetic energy
to heat energy

The cyclist begins with energy from food which is changed to kinetic energy as he is cycling and ends as heat when he applies the brakes. However the total *amount* of energy remains the same even though the *type* of energy changes. If the cyclist gained 1000 J (joules) from his food he might convert 250 J to movement energy. The other 750 J would be changed to other forms of energy such as heat and sound, and some might be stored.

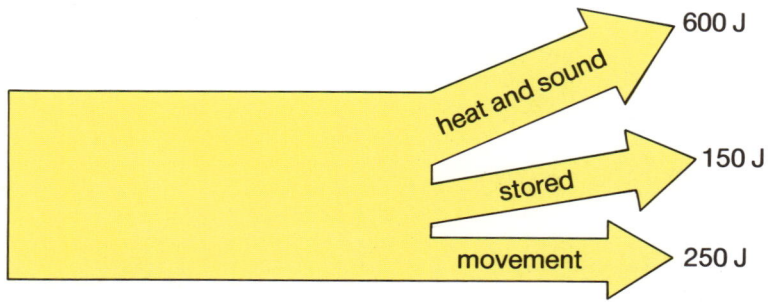

heat and sound — 600 J

stored — 150 J

movement — 250 J

The total amount of energy remains the same

Call my bluff

The class should divide into groups. Each group will be asked to work on one of the following statements.

Prepare evidence which you can use to convince others that your statement is true. You can use real information or make up your own.

Each group must present its evidence to the rest of the class and a vote will be taken to decide which is the correct statement.

1 The law of conservation of energy is a law passed by Parliament which says you must save energy.
2 The law of conservation of energy is a law which says you should try not to spoil the countryside, and must try to keep plants and animals alive.
3 The law of conservation of energy is a law which says that if you have a supply of energy, you cannot use all of it to do what you want.
4 The law of conservation of energy is a law which says you can change energy from one type to another but the total amount stays the same.
5 The law of conservation of energy is a law which says you can use energy only when necessary.

Q1 Which statement gained the most votes? Was this the correct one?

Q2 The other statements have an element of truth in them. For each one explain what this is.

 Take notes

Discuss with a partner and write down:

- what energy is used for
- what types of energy there are
- an example of energy changing from one type to another.

Input and output

Energy changers, or **transducers**, change energy from one type to another. You are a transducer because you change chemical energy to heat and movement. An electric light bulb is also a transducer as it changes electrical energy to light and heat.

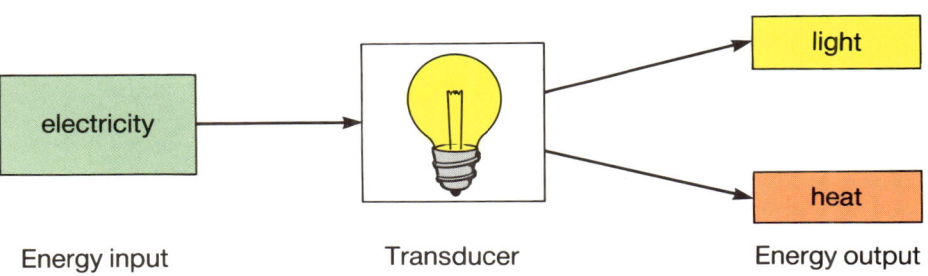

Energy input Transducer Energy output

Go round the laboratory and identify the energy input and the energy output for the transducers on display.

Record in your notebook, by drawing diagrams like the one on the previous page.

Compare notes with others in your group. Do you all have the same answers? Do you think there may be more than one correct answer in some cases?

Q1 What was the most common energy output?

Q2 What proportion of transducers had more than one form of output?

Q3 a Select a room at home and make a list of energy transducers in it.

b Compare your list with others in your class. Which room contains most energy changers?

c In the event of an electrical power cut which energy changers from your list could you still use?

Q4 The table shows when the first practical versions of transducers were produced.

a Copy the table into your book, leaving space between each listed item. Fill in the **energy changed** columns. The first has been completed for you.

Year	Transducer	Energy changed from	to
1800	Electric battery	chemical	electrical
1820	Camera		
1831	Electric generator		
1840	Bicycle		
1859	Internal combustion engine		
1876	Microphone		
1878	Filament light bulb		
1884	Steam turbine		
1896	Radio set		
1902	Photocell		
1926	Television		
1945	Radio telescope		
1982	Compact disc player		

An early 'Boneshaker' bicycle

An early camera used to take daguerreotypes

The first internal gas combustion engine

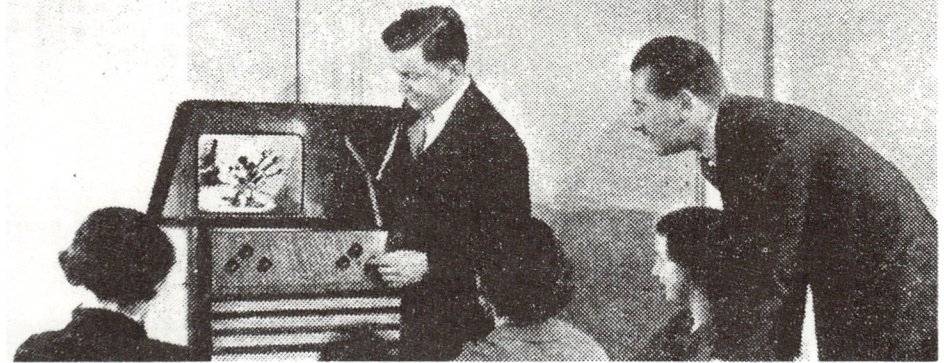

The first television pictures

An early radio set

b Many modern energy changers involve electricity. Use the resources either in the classroom or the library to research transducers which do not use electricity and add them to your table.

c Many of these energy changers were developed to make life easier or more enjoyable. Make lists to show which make life easier and which make life more enjoyable. Are there any you could do without?

d Select two energy changers and draw a diagram for each to follow the energy back to the sun.

Project

Find out what machines were used by an ancient civilisation, e.g. Romans, Greeks, Chinese, Egyptians. Produce a report for the class. Include diagrams and a list of where the information was obtained from. Perhaps a poster could be included.

Energy resources

Fire to furnace

The United Kingdom is an industrialised country which uses vast amounts of energy.

 5% other

 10% transport

 15% heating

 70% manufacturing

The main uses of energy in the United Kingdom

The control of energy from fuels was one factor that helped humans to make material progress. People began with simple fires but now we use a wide variety of fuels. If there is a crisis such as a power cut we soon see that almost every part of our lives is affected. Unfortunately we cannot rely upon fuels for ever since the world's resources of fuels are being used up.

Some energy sources are **renewable** but some are not. Plants which are grown for food are renewable but it takes millions of years for plant remains to be turned into coal. It is also likely that in the conditions that exist now coal is not being made at all. Coal is being used up; it is **non-renewable**. Natural gas and oil are also non-renewable. Oil is extremely useful because it provides petrol and diesel and it is also the raw material from which plastics are made. Unless new sources of oil are found and exploited, the supply will run out early next century. Can you imagine a life without plastics?

Coal, oil and gas are called **fossil fuels** because they are made from the remains of living things. The energy locked in fossil fuels originally came from the sun. Fossil fuels and nuclear fuel are 'primary' fuels from which electricity is produced.

Electricity can also be produced from water power in a hydroelectric dam. This must be built where the conditions are right, e.g. in a mountainous area with high rainfall. Hydroelectric power is a renewable energy source.

The sun is the source of all non-nuclear energy

A hydroelectric dam

It makes sense to investigate the possibilities of using renewable energy sources and there are several alternatives now in use across the world.

Wind turbines

A tidal power station

Solar panels

A nuclear power station

Geyser steam units at a geothermal power station

Alternative energy sources

1 Collect a cut-out sheet and try to match the energy source with the 'for and against' arguments.
2 Stick the correctly matched information in your book.
3 Discuss which energy source is the odd one out and why. (**Hint**: are all the sources strictly renewable?)

Q1 Which sources would be suitable for development in the UK? Give reasons for your choices.

Q2 Make a pie chart to show the use of energy in the UK.

[IT] Use a database to investigate the relative costs and distribution of the main energy sources in the UK.

 *Take notes*

Discuss in a group and write down:

- the main ways energy is used in the UK
- a table showing sources of non-renewable energy.

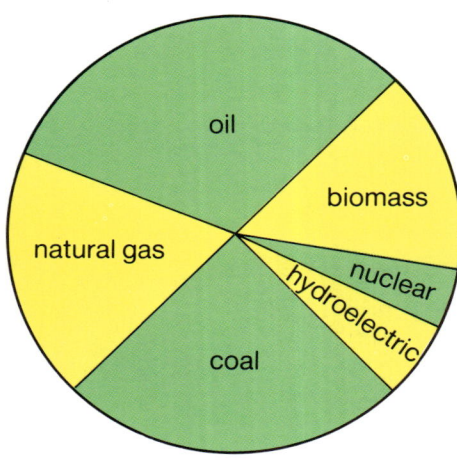

World energy sources in 1990

Biomass

Fuel stocks are limited. We need to do two things urgently:

- Use the remaining stocks of fossil fuels more efficiently.
- Develop alternative efficient sources of energy.

The most efficient way of trapping the sun's energy is by growing plants. **Biomass** is organic material which depends on the activity of plants and photosynthesis. It includes plants and animals (including us) and their waste products.

China has over 7 million digesters which produce methane gas from biomass waste. Methane gas can be used as a fuel. In Britain much of the 20 million tonnes of rubbish thrown away each year is biomass and this could be used to provide energy. Some local authorities are exploring this idea. Biomass now provides a significant proportion of the world's energy.

Wood chips being loaded into a reactor which converts the biomass into methane-rich gas by treating with steam in the presence of a chemical catalyst

Energy from biomass

1 Set up the two vacuum flasks as shown. Loosely fill with grass or nylon. Record the initial temperature and then check it daily until change is seen. Produce a report.

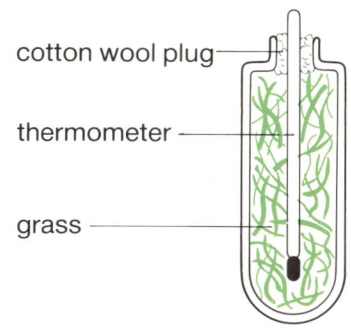
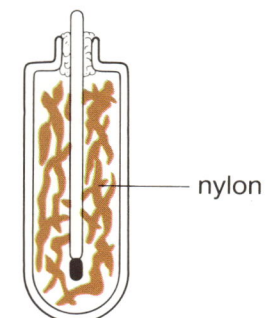

cotton wool plug

thermometer

grass

nylon

2 How could you use the energy from a compost heap to keep plants warm in a greenhouse or coldframe?

- Carry out research by looking for information in books and magazines, at garden centres, or by asking a gardener you know.
- Draw a diagram to show how your idea will work.
- If you can do it in a laboratory, ask your teacher for the equipment and test your design.
- Produce a report on your findings.

CHECKPOINT

📚 Project

Find out what your local authority is doing with its waste. Produce a report.

Gravitational fields and weight

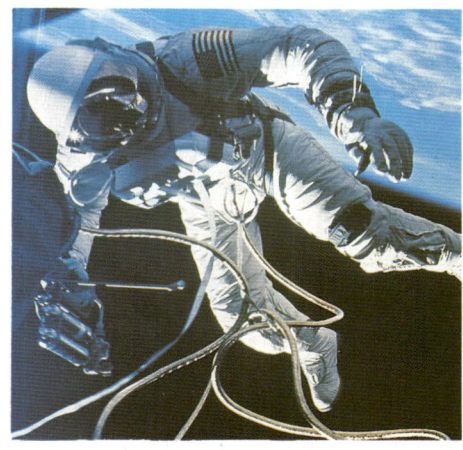

What goes up must come down?

Most objects do fall back to Earth, because of one of the basic forces in the Universe. This is the attraction between two masses or lumps of matter which we call the **force of gravity**.

If you pick up an object you can feel the gravitational attraction of the Earth and this pull or force on the object is called the **weight**. The force of gravity acts on objects all around the Earth and is the effect of a **gravitational field**. On the Earth's surface the gravitational field pulls with a force of 9.8 newtons on every kilogram of matter.

gravitational field strength g = 9.8 N/kg on the
Earth's surface

Mass and **weight** are often confused. The mass of an object is the amount of matter it contains and it is measured in kilograms. The weight of an object is the force on it due to gravity. The unit for weight is **newtons**, not kilograms as used in everyday life. If you went to the moon your mass would be the same as on Earth but you would weigh only one-sixth of your weight on Earth. This is because the gravitational field strength of the moon is one-sixth of that of the Earth.

The gravitational field of the Earth keeps satellites in orbit. If there were no pull of gravity the satellite would continue in a straight line after being launched and would soon be lost in space.

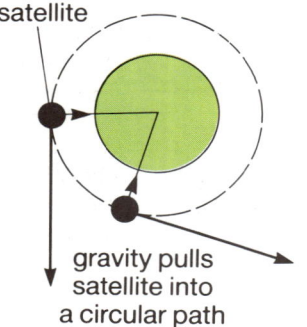

satellite

gravity pulls satellite into a circular path

The Earth and moon attract each other and this gravitational field causes the moon to orbit the Earth. (It also causes the ocean tides.)

Sir Isaac Newton tried to explain why gravity causes the moon to orbit the Earth by imagining a high mountain with a large gun on it. This gun fires a shell (projectile) horizontally from the top of the mountain but because of its speed and the curve of the Earth the shell goes a very long way before falling to the ground.

Newton thought that if there were a mountain much taller than the atmosphere then a shell with sufficient speed would continue to fall towards the Earth, but because of the curve of the Earth's surface the shell would never actually reach it. It would remain at the same height above the ground. The shell would then be a satellite in orbit around the Earth.

Q1 An astronaut has a mass of 66 kg on Earth.
   a What is his mass on the moon?
   b What is his approximate weight on the moon?

Q2 The table below shows the mass of the planets compared with the mass of the Earth and the pull of gravity on a one kilogram mass (the gravitational field strength) near to their surfaces.

Planet	Relative mass	Gravitational field strength (N/kg)
Mercury	0.06	2.8
Venus	0.82	8.9
Earth	1.00	9.8
Mars	0.11	3.9
Jupiter	318	25
Saturn	95	10.9
Uranus	15	11.0
Neptune	17	10.6
Pluto	0.002	2.8

a On which planet would you feel lightest?
b What do you think would be the long-term effects on the body's muscles if you lived on a planet with gravity much greater than on Earth?

c Plot a graph of planet mass against gravitational field strength.

[IT] Use a data-handling program to plot the graph.

d Comment on the shape of your graph. What is the relationship between the planet mass and the field strength?

e Draw a table to show the weight of the astronaut from question 1 on each of the planets.

Pairs of forces

Newton hypothesised that if object A exerts a force on object B, then object B exerts a force back which is of equal size but opposite in direction.

Measuring forces

A force can be used to stretch a piece of elastic or a spring and this can be used to measure the size of the force.

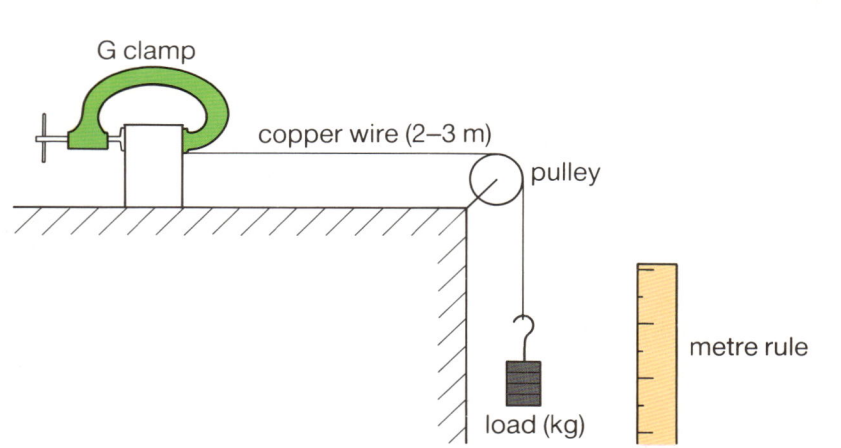

1 Your teacher will demonstrate what happens when copper wire is stretched.

- Record the load on the wire and the amount of stretch (extension).
- Plot a graph with load on the vertical axis and extension on the horizontal axis.

Collect
- masses
- G clamp
- pulley
- springs
- rubber band
- nylon thread
- metre rule

2 Work in groups.

- Repeat the experiment using a spring, a rubber band, and nylon thread.
- Plot load–extension graphs for each.

3 Compare the four graphs and comment on their shapes.

Q1 Look at the pictures of pairs of forces on the opposite page. Draw a table to show the opposing forces.

Q2 Explain how one of the Newton-meters shown opposite works.

Q3 When a material does not go back to its original length or shape after stretching it is said to have exceeded its elastic limit. Do you think any of the materials in your investigation exceeded their elastic limit? What would happen if they did?

Take notes

Discuss in a group then write down:

- what gravity is
- the difference between mass and weight
- what keeps the moon in orbit
- one way of measuring force.

Stretching problems

Work in pairs and choose one of the following problems.

Show your design to your teacher and ask for the equipment you require for next lesson.

1 Design your own forcemeter using what you have learnt about stretching. You could start with some cardboard and a spring. What else will you need?
2 Design an experiment to find out which elastic keeps its elasticity best after repeated washing.

Project

Write to an elastic manufacturer and ask what tests are carried out on their elastic. If possible use a wordprocessor to write and edit your letter.

Q1 Compare your results with those of other pairs.

Q2 Did your design work? Could you improve it?

It is much easier to push-start a supermarket trolley when it is empty. It is much harder to stop it when it is full

Stay as you are

We have said that **mass** is the amount of matter in an object. Another definition is that mass is a measure of **inertia**. Inertia is from the Latin for laziness. What are the effects of inertia in the cartoons a–c above?

Inertia is the tendency of an object to stay at rest or to keep moving at constant speed in a straight line unless a force acts upon it.

It is difficult to move a large mass but once it is moving it is difficult to stop. The larger the mass, the greater the inertia.

Collect

- ramp
- trolleys
- metre rule
- balance
- hacksaw blade
- G clamp
- plasticine
- stopclock
- blocks
- protractor

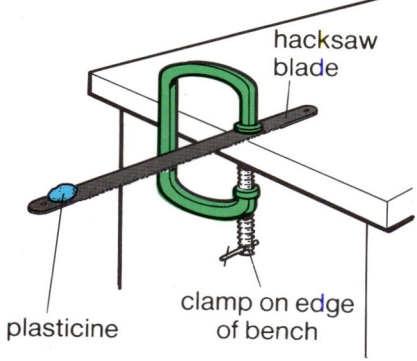

Exploring inertia

1 Using trolleys of different masses, investigate the relationship between the mass of the trolley and the distance from the bottom of the ramp the trolley travels after it runs down the ramp.

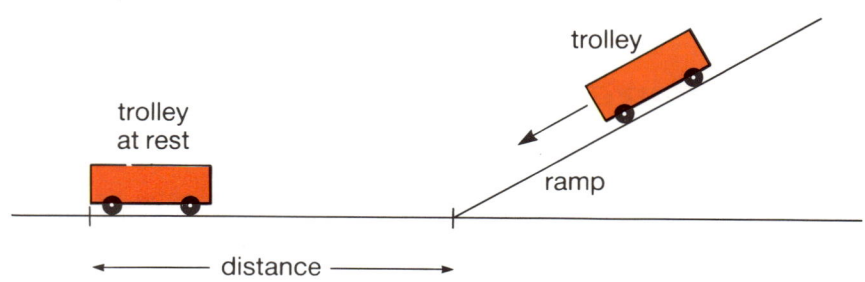

2 Clamp a hacksaw blade onto the end of the bench. Attach a 50 g lump of plasticine and start the hacksaw vibrating. Count the numbers of vibrations in 30 seconds.

Repeat with different masses of plasticine. Investigate the relationship between the mass and the numbers of vibrations. Record your results.

To change the speed of an object a **force** must act on it. The larger the mass of the object the greater the force needed. Also, the greater the change in speed, the greater the force needed.

It is common sense to think that the harder you push something, the faster it goes. Everyone has discovered this when playing with toy cars or pushing supermarket trolleys.

Newton put this more scientifically in his Second Law of Motion:

F = force in newtons
m = mass in kilograms
a = acceleration in m/s^2

The acceleration of a body (the change in speed per second) is directly proportional to the force causing the acceleration.

$$F = ma \quad \text{or} \quad a = \frac{F}{m}$$

The following pictures illustrate this law.

If the force or push is increased and/or the mass decreased then more acceleration is possible

Stability

Every mass on Earth has a gravitational force acting on it which we call its weight. The point in an object where the force of its weight seems to act is called its **centre of gravity**. This coincides with where the object's inertia seems to be concentrated, the **centre of mass**.

If a body does not topple over easily, it is described as **stable**; an **unstable** body topples easily. Consider the two positions of the box in the diagram. Which do you think will topple more easily if pushed near the top?

There are two ways to make an object more stable:

a lower the centre of gravity
b widen the base line.

Exploring stability

Vehicles which are likely to become top heavy are tested for stability

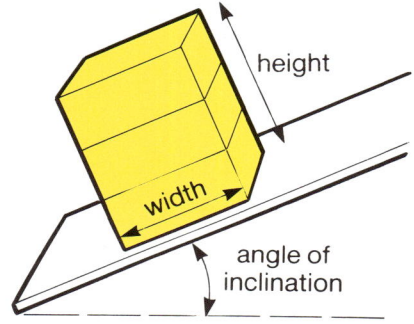

Collect
- blocks
- protractor
- ruler

Using only blocks, a ruler and a protractor, investigate:

a how the height and width affect the stability of the blocks
b how the angle of inclination affects the stability of the blocks.

Q1 Write a report about your investigations.

Q2 What can you say about the conclusions in each case?

Q3 Do you think the investigations involved fair comparisons? Are there things you did not take into account which might have affected the results? (Did the trolleys all have similar tyres, for example?)

 **Take notes**

Discuss in a group and write down:

- two definitions for mass
- an example of inertia
- a definition of weight
- two ways to make an object more stable.

Balancing acts

- cardboard box
- lead weight
- cardboard

- 2 candles
- 2 glasses
- cork
- long pins
- long knitting needles

- card
- scissors
- plasticine
- weight
- clamp
- clamp stand
- thread

A magic box

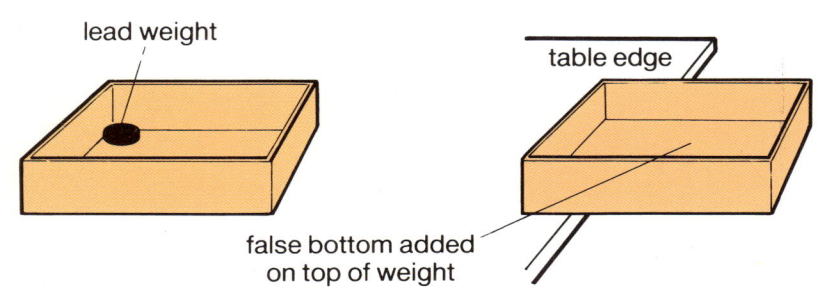

lead weight

table edge

false bottom added on top of weight

Candle seesaw

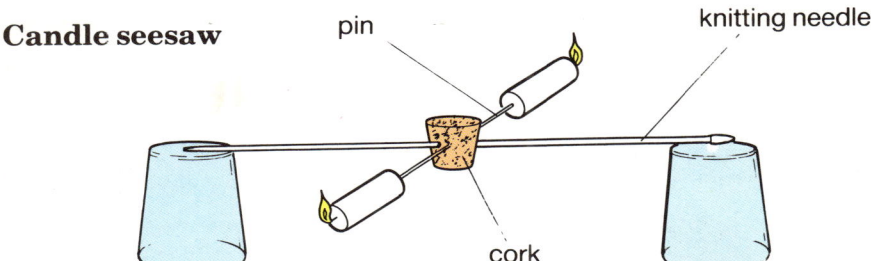

pin

knitting needle

cork

Swinging parrot

Make a parrot as in the picture and add a card stand as shown. Add some plasticine to the tail and balance the parrot on its stand on a shelf. To find the best position for the stand you need to find the centre of mass. Suspend the parrot from several points by thread and attach a plumb line. Mark along the plumb line. The point where the lines cross is the centre of mass.

stand

Q1 Explain why each of the balancing acts works, in terms of the centre of mass.

Q2 Draw a diagram to show where the centre of mass of the parrot is and how you found it.

Project

Draw a design for a stable high-chair.

Energy, power and work

A matter of definition

Energy, **power** and **work** are used in everyday language to mean a variety of things. In science, however, they have quite specific definitions.

Energy is a measure of the ability to do work (in joules).

Power is the rate of doing work (in joules per second, or watts).

Doing work . . .

. . . and doing the same work faster

If you are sitting in a chair watching television you have lots of stored chemical energy from the food you have eaten, but the only work you are doing is that required for normal bodily functions. When you run upstairs you are doing more work and using more energy to do it. The amount of work you do depends on your weight and the height of the stairs. Your friend may be able to run up the stairs faster than you. If you are the same weight you would say that he is more powerful than you because the time he takes to do the work is shorter. His **power output** is greater.

Efficiency

As you have seen earlier in this unit, energy cannot be destroyed but it can be changed to other forms. When you run upstairs some of your energy is changed to heat and you may sweat to cool down. A car converts much of its energy to heat, most of which is not wanted, so it also needs a cooling system otherwise the excess heat would damage the engine. When heat energy is lost from a machine the power output is considerably less than the power input. The machine is less than 100% efficient. No machine can ever be 100% efficient.

Efficiency means different things to different people. To the railway user it may mean that trains run on time; to the railway manager it may mean that the railway runs without losing money; to the scientist and engineer, it means the **effective use of energy**.

Power values can be used to find out the efficiency of a machine or engine (including the human engine). The answer is usually shown as a percentage.

$$\text{efficiency} = \frac{\text{power output}}{\text{power input}} \times 100\%$$

An efficient railway?

Measure your power output

Work in groups.

Collect

- scales
- metre rule
- stopwatch
- sand bags

1 a Measure your weight in newtons. (Remember that 1 kg weighs approximately 10 newtons.)

b Measure the vertical height of the stairs in metres.

$$\underset{\text{(joules)}}{\text{work done}} = \underset{\text{(newtons)}}{\text{force (your weight)}} \times \underset{\text{(metres)}}{\text{distance (height)}}$$

The rest of the group should measure the time you take to run up the stairs. Calculate your power output.

$$\underset{\text{(watts)}}{\text{power}} = \frac{\text{work done (joules)}}{\text{time taken (seconds)}}$$

2 Use the above ideas to measure your power output by

- stepping up and down off a stool
- lifting sand bags
- lifting books
- opening a door.

Q1 What was your power output?

Q2 Draw a bar chart to show the power outputs of each member of the class.

Q3 Who had the greatest power? Was this what you expected? If not, why not?

Measure the efficiency of a simple pulley

Set up the pulleys as shown below.

Collect

- 2 pulleys
- clamps
- stands
- masses

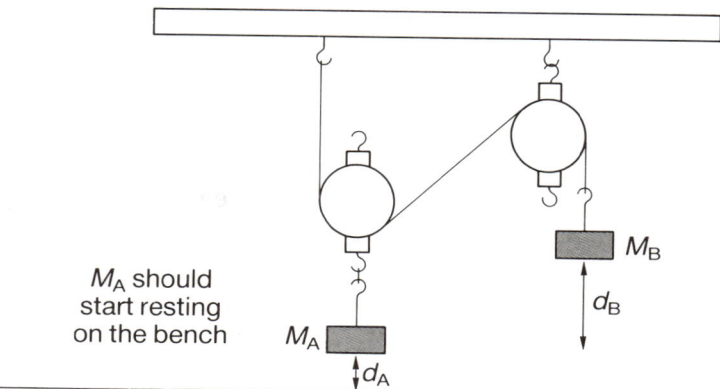

M_A should start resting on the bench

M_A

d_A

M_B

d_B

Hang a load from pulley A then hang just enough masses from pulley B to move the load attached to A.

1 Measure the distance moved by both loads.
2 Calculate the weight in newtons of the masses (assuming 1 kg weighs 10 N).
3 Calculate the work done by each load.
4 Calculate the efficiency of the machine.

$$\text{efficiency} = \frac{\text{work output (A)}}{\text{work input (B)}}$$

⟵

(Note: This equation can be used for pulleys because it takes the same time for each side to move.

$$\text{power} = \frac{\text{work}}{\text{time}}$$

If time is removed from the equation then work can be substituted for power when working out the efficiency of the above pulley.)

Design a table to record your results and repeat using loads of different masses.

What happens if you use three pulleys?

Q1 Write a description to show how you measured the efficiency of your pulley system.

Q2 Did changing the load effect the efficiency?

 **Take notes**

Discuss in a group and write down:

- a definition of energy and power
- an energy or block diagram showing the energy changes after someone has run upstairs
- an example showing why no machine is 100% efficient
- a definition or equation for efficiency.

CHECKPOINT

120

Bags of joules

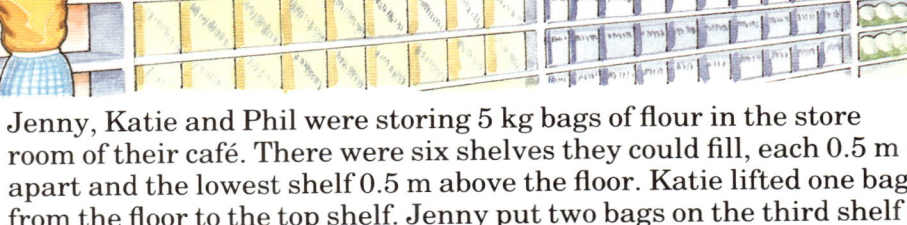

Jenny, Katie and Phil were storing 5 kg bags of flour in the store room of their café. There were six shelves they could fill, each 0.5 m apart and the lowest shelf 0.5 m above the floor. Katie lifted one bag from the floor to the top shelf. Jenny put two bags on the third shelf and Phil put three bags on the second shelf. Katie then said she had done the most work. Was she correct?

Remember: work done = force × distance moved

Complete the table to discover if Katie is right.

Weight of bags N	Height of shelf m	Work done J

Q1 A delivery is made of another 45 kg of flour. Which shelves should they put them on so that they all do the same amount of work? (No shelf can hold more than three bags.) How many ways of doing this can you find? If necessary collect a resource sheet to help you to work this out.

Q2 Using the information from the table work out how much energy is transferred when a parcel is moved from shelf three to shelf four.

Q3 How much less energy would a parcel have if you moved it from the top to the bottom shelf?

As you know distance is measured in metres, force in newtons and work (and energy) is measured in joules. Both force and work are named after scientists. Joules are named after James Prescott Joule who lived in Sale near Manchester about 140 years ago. The house he lived in is still standing and it has a plaque on the outside to commemorate his living there.

*Remember
1 kg weight = 10 N*

📚 *Project*

Find out more information about James Joule and write a newspaper article about his contribution to science. Include historical and personal notes to make it interesting.

Problem

Eggsistance

For every force there is an equal and opposite force. When an egg is dropped the force of the egg's falling weight acts on the floor. The floor exerts an equal force on the egg, with obvious results!

When a force acts on an object it may

- change its speed, and/or
- change its direction, and/or
- change its shape.

Remember:

$$\text{force} = \text{mass} \times \text{acceleration}$$
$$= \text{mass} \times \text{rate of change of speed}$$

When the egg reaches the floor its speed is reduced to zero in a short time and the force is large enough to break the egg.

What happens if you drop an egg onto a pillow? Why?

Your task is to design a container made from paper which will protect a raw egg when dropped from a height of one metre onto a hard floor.

- Draft your own design for the container then discuss with your group whose is the best design. What makes it the best design?
- In a group, build the container using your chosen design and test it.

How will you test it? You don't want to break dozens of eggs!

- Prepare a report on your design, your tests and your results. Explain why you chose that design.
- Could you improve on the design?

Car safety

Nowadays cars are tested thoroughly for safety. Some cars have crumple zones around the passenger areas to absorb some of the car's kinetic energy. They also have a rigid section on the sides especially to protect passengers. Look through this chapter and see if you can think of other ideas for car safety.

Q1 In your group discuss the features labelled in the diagram below.

    a Which of these are compulsory by law?

    b Safety in a car, as elsewhere, is a compromise between design, cost, space and other factors. If the manufacturer could only afford to put six of these features into effect, which would you choose and why?

    c Are there any other improvements that you can think of which will assist with safety? How expensive do you think it would be to put into effect?

    d It has recently been suggested that the speed limit on motorways should be increased from 70 mph to 80 mph. If this is put into effect how would the government compare safety aspects?

Q2 Magazines such as *Which?* and *What Car?* give information and specifications on cars. Examine some up-to-date copies and compare the safety features incorporated into some of the popular models.

Q3 Prepare an advertisement to sell a car. Set up a panel to ask questions and award points for safety. Which is the safest car from those selected?

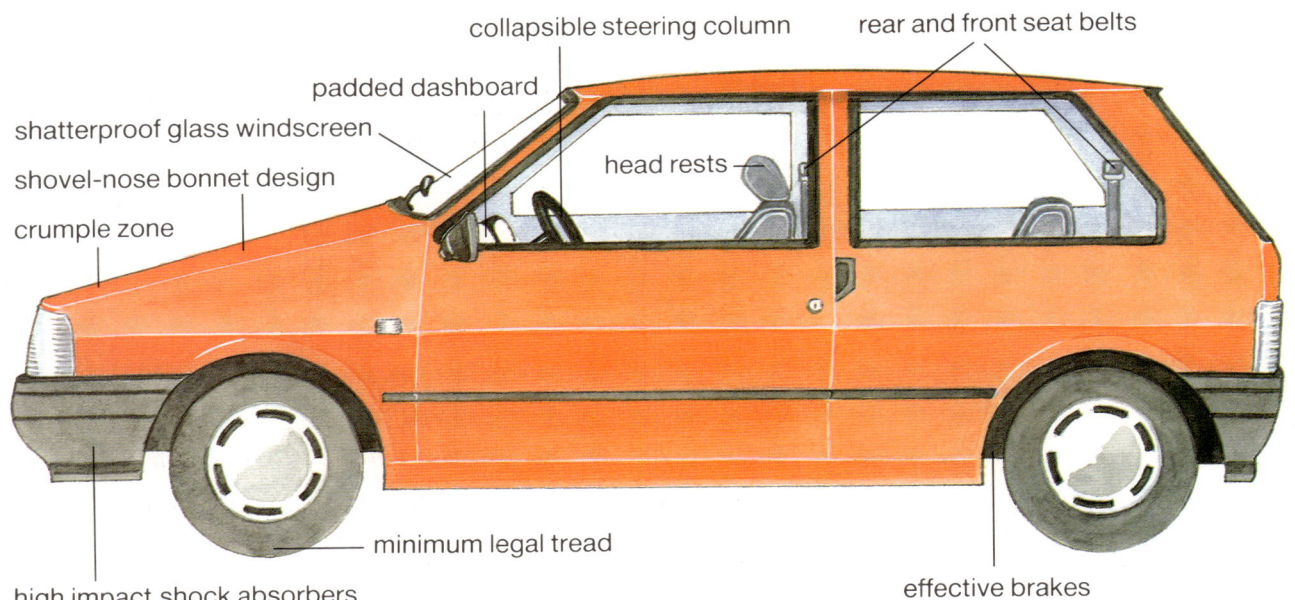

collapsible steering column — rear and front seat belts

padded dashboard

shatterproof glass windscreen

shovel-nose bonnet design — head rests

crumple zone

minimum legal tread

high impact shock absorbers — effective brakes

1 a Make a sketched copy of the diagram and draw an arrow to show the *direction* in which the man's weight acts. [1]

b His mass is 62 kg. What is his weight in newtons? [1]

MEG 1988 Physics (Nuffield)

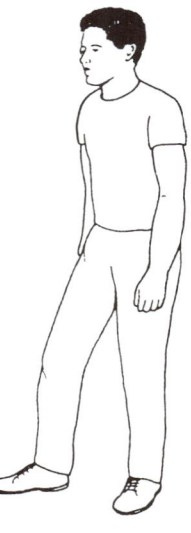

2 A trolley is pulled back against an elastic band, as shown in the diagram.

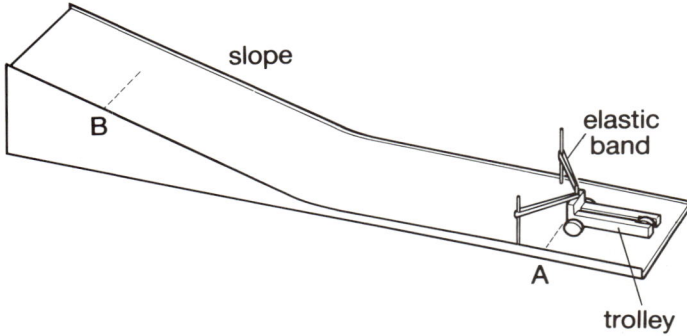

slope

B

elastic band

A

trolley

It is then released from position **A**. The trolley runs along the flat surface and then up the slope to **B**.

a The trolley comes to rest at **B**. What type of energy does it have at **B**? [1]

b The trolley then runs back from **B**. It is stopped briefly by the elastic band stretching again. Using axes as shown below, sketch a graph of *speed* against *time* for the trolley *from the moment it leaves* **B** until it comes to rest again. [3]

MEG 1988 Physics (Nuffield)

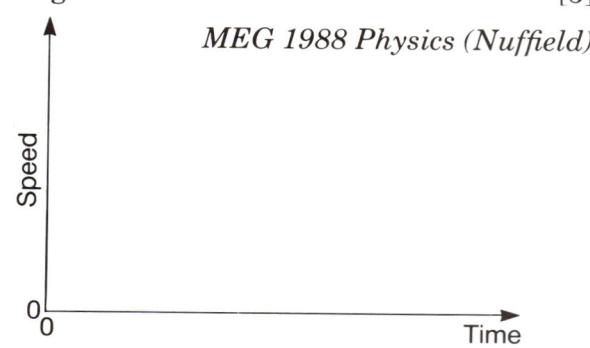

3 A scientist did an experiment to find how a steel spring stretches with the load applied. He got the following readings.

10 − 4   20 − 12.

Load newtons	Extension millimetres	Load newtons	Extension millimetres
1.0	1.3	4.0	5.3
2.0	2.7	5.0	7.0
2.5	3.3	6.0	8.0
3.0	4.0	8.0	10.7
3.5	4.7		

a On a piece of graph paper plot a graph with the extension along the vertical axis and the load along the horizontal axis. [6]

b Use the points to draw what you think is a suitable line to show how the spring behaves when stretched. [2]

c What does the slope of the graph tell you about the spring? [2]

MEG 1988 Physics Specimen

4 A test was carried out in which a car driver applied her brakes to bring the car to a halt, without skidding. The distance the car took to stop (its *braking distance*) was measured for different *speeds*. The results were:

¹⁄₁₀ ¹⁄₅ ¹⁄₃.   ¹⁄₂.

Speed of car kilometres/hour	10	20	40	60
Braking distance metres	1	4	16	36

a Why is the braking distance greater at higher speeds? [1]

b The same test was carried out with three extra passengers in the car. The braking distances were longer. Explain this. [2]

c Where does the car's kinetic energy go when the brakes are being applied? [2]

d Complete the table by writing down a likely value for the braking distance (without passengers), for a speed of 60 kilometres/hour. [1]
Give a reason for your answer. [1]

MEG 1988 Physics (Nuffield)

7
Regulation and control

Body temperature

Reminders for using thermometers:

- Check that you can read the thermometer correctly.
- Clinical thermometers need to be shaken to get the mercury down.
- Disinfect your clinical thermometer after use.

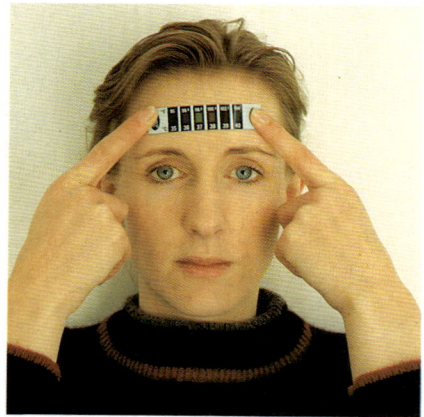

Liquid crystal thermometer

Part of body	Type of thermometer used
forehead	LCT
under the tongue	clinical thermometer
armpit	clinical thermometer
forearm	LCT
back of hand	LCT
clenched palm	clinical thermometer
bend of knee	clinical thermometer
calf	clinical thermometer
top of foot	LCT

Work in pairs. One of you should sit down and be the subject, the other takes the measurements. Note everything that you find out.

[IT] A computer thermometer probe could be used.

Discuss these points with your partner.

1 What is the range of temperatures? (The highest and lowest and the difference between them.)
2 How do your results compare with those from other groups? Which temperatures are most alike?
3 Sort your measurements into two sets: set A higher temperatures and set B lower temperatures. Explain your groupings.
4 Which parts of the body are good for measuring temperature? Why?

Q1 Report on what you discovered and make sure you cover all the points that you and your partner discussed.

Q2 Work out the class average for mouth temperature.

[IT] Use a spreadsheet program to collate the class results.

Core temperature and shell temperature

The photographs on the next page were taken by a special technique called infrared thermography.

The warmer parts of the body are shown in red and the cooler parts in blue. Body extremities such as hands, feet, arms and legs (**shell**) vary in temperature much more than the **core** (deeper) parts of the head and body. If you go outside in winter without gloves you soon notice that your hands will get cold.

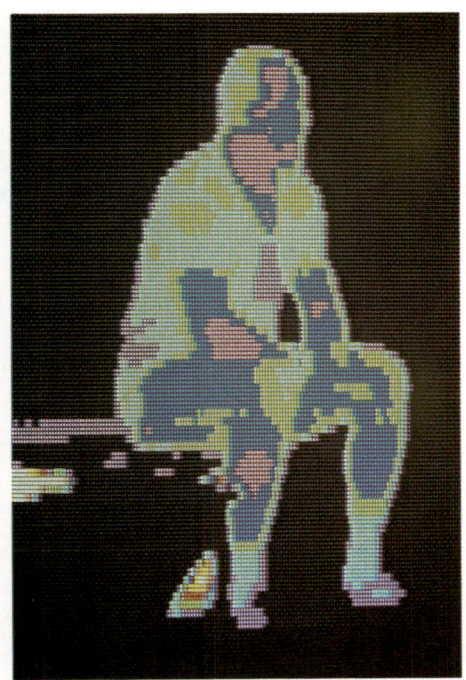

Thermogram of the body showing temperature variation, from white (hottest) through yellow, red, blue, green, purple, to black (coldest)

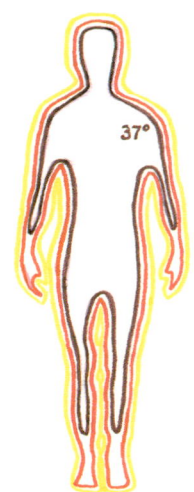 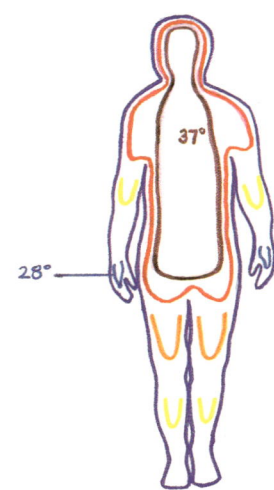

37° 37°

28°

Body isotherms (lines of equal temperature) in hot (*left*) and cold (*right*) weather

 *Take notes*

Discuss with your partner and write down the meaning of core temperature and shell temperature.

CHECKPOINT

Hot and bothered

See what happens to the temperature of different parts of your body when you exercise. You could try running on the spot, stepping on and off a wooden block or, if it is convenient, running round the playground. Observe all the changes that take place.

1 Ask your partner to take the temperature of some different parts of your body before you start. Good ones to do are mouth, armpit, forehead, clenched palm. Use a liquid crystal thermometer or a clinical thermometer, as appropriate.

[IT] A computer thermometer probe could be used.

2 Keep exercising for as long as you can (at least two minutes). Time how long you exercised.
3 Get your partner to take the temperature of the same parts of your body as soon as you stop your exercises.
4 Press a piece of anhydrous (blue) cobalt chloride paper in the palm of your hand and another on your forehead.

Q1 Write a full report of your experiment.

Q2 Compare the temperatures you found after exercise with temperatures taken before exercise when sitting normally.

Thermogram showing body temperature after the person has been playing squash

Q3 Suggest how your body caused these temperature changes.

Q4 Why is it important for skin temperature to be able to change?

Q5 What does the anhydrous cobalt chloride paper test show?

How do our bodies produce heat?

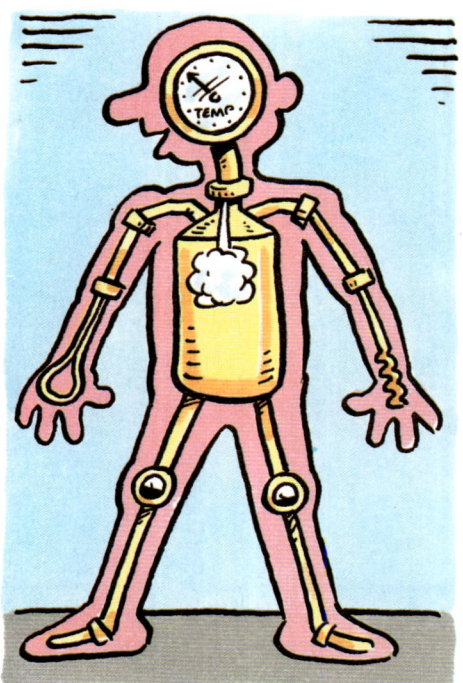

All the cells in our body are respiring and releasing energy from food, e.g. glucose (see p. 139). This is part of the set of processes called **metabolism**. The speed of metabolism when we are at rest is the **basal metabolic rate**. This governs how quickly food is broken down and also how quickly energy is released. This energy might be used by the cell in a variety of ways like muscle movement or growth and repair of cells. Some of the energy is not put to work by the cell but is released as heat. Muscle cells obviously produce movement but they also generate a lot of heat. The liver also produces heat.

In humans, normal core temperature is usually 37.5 °C. The temperature under the tongue is more likely to be 37 °C. There are slight variations from person to person and also slight daily variations but, unless you are ill, your temperature is not likely to have a range of more than 1.5 °C. This core temperature is the same whether you live at the South Pole or the Equator!

For our bodies to function efficiently we need to keep our internal environment as constant as possible. The name given to this process is **homeostasis**. Regulating our body temperature is an important part of this process.

Do all animals control their own temperature?

Some animals produce their own heat and control their temperature from inside their own bodies. These organisms are called **endotherms**. Other animals get practically all their body heat from outside their bodies. These are called **ectotherms**.

Fish, amphibians and reptiles are ectotherms

Birds and mammals are endotherms

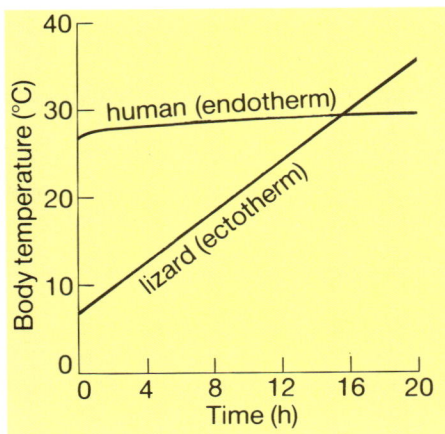

Endotherms' body temperature remains constant while ectotherms' body temperature may vary

Some ectotherms have a temperature which varies because the temperature of their surroundings varies. For example, desert lizards and snakes will be cold at night when the temperature of the air is very low and hot during the day in the sunshine.

Living cells function best at temperatures between 30 °C and 40 °C. This is called the **optimum temperature**. **Enzymes** (catalysts that speed up chemical reactions in the cell (p. 59)) are very sensitive to changes in temperature and they work most efficiently over this range.

The rates of reactions in cells increase or decrease as the cells' temperature increases or decreases. If the cells are too cold then the enzymes will work more slowly and metabolism will be very slow. So, ectothermic animals tend to be very sluggish at night and early in the morning and could easily end up as some other animal's breakfast! As the temperature increases during the day they become more active. If the cells become too hot, the enzymes are **denatured** (destroyed). The animals have to find shade or some other way to cool down. Their behaviour at different times of day helps to control their temperature. Otherwise they could seriously overheat and might die. Endothermic animals are less likely to die from overheating or cold but they do need a constant supply of food in order to generate heat.

The ectothermic lizard uses energy from the sun to increase its core temperature

Take notes

Discuss with your partner and write down two or three (no more!) key sentences under each sub-heading. Drawing diagrams can help you to summarise and remember facts.

The endothermic sheep needs a constant supply of food to maintain its core temperature

Project

Find out what advice is given to elderly people about keeping warm in winter. What scientific ideas are used to prepare this advice?

Regulating temperature

Steady state

Endothermic animals must keep their core temperature between quite narrow limits. If the core temperature of a human being goes up or down by as much as 4 °C then he or she is in serious danger. In temperate (cool) places like Great Britain the temperature of the surroundings is lower than our optimum core temperature. In these conditions, endotherms are constantly losing heat.

Hot objects, including animals, lose heat to their surroundings by three processes:

- Conduction—occurs when a body is touching something colder than itself. Heat flows from the hot body to the cooler body.

- Convection—occurs when moving air or water is warmed by something hotter and the warmed air or water is carried away in a current. Colder air or water moves in to replace the warmed air or water.

- Radiation—involves the loss of heat energy by electromagnetic waves. Heat waves can travel through space or air.

If an animal's skin is wet, heat can be lost even more quickly. After swimming you feel colder until all the water on your skin has dried up. This is because heat energy is being taken from your body to turn the water into a vapour. This is known as **evaporation**.

Collect

- 2 large test tubes
- 3 beakers
- 2 thermometers
- ice
- bunsen burner
- heat-proof mat
- tripod and gauze
- safety glasses

What influences the rate of cooling?

1 Set up this experiment. Use a beaker to heat the water for the test tubes.

2 Take the temperature in each test tube every minute until there is no further change in the temperature. Stir to even up the temperature.

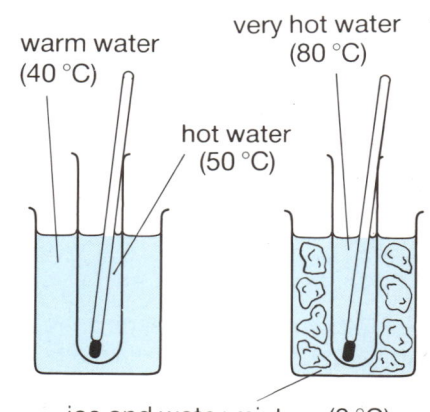

warm water (40 °C)

very hot water (80 °C)

hot water (50 °C)

ice and water mixture (0 °C)

Q1 Write a report on your experiment. Include two graphs drawn on the same axes.

Q2 Explain what this experiment shows you about heat loss and the difference in temperature between a hot object and its surroundings.

Q3 Think of a polar bear and an elephant. Which might lose heat faster?

Collect

- large can
- small can
- 2 tripods
- large beaker
- 2 thermometers
- bunsen burner
- heat-proof mat
- safety glasses
- graph paper

Amount and rate of cooling

1 Take a big can and a small can and rinse them with very hot water. Each can should have a lid with a hole for the thermometer.

2 Balance them on the tripods.
3 Very carefully, fill them with water that is almost boiling. Replace the lids.
4 Take the temperature of each can of water—they should be the same (or very nearly).
5 Take the temperature every two minutes for twenty minutes.

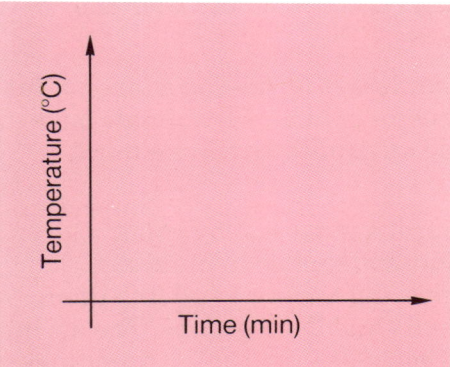

Q1 Write a report of your experiment. Include
- a diagram
- a table of your results
- a graph of both sets of results on the same axes
- your conclusion.

[IT] Use a data-handling program to plot the graphs.

Q2 Think of a mouse and an elephant. Which do you think has more difficulty keeping its temperature up? Which will have to eat more often?

Why don't humans (and other endotherms) cool down to room temperature?

Endotherms manage to keep up their temperature in two major ways.
1 They reduce heat loss to the surroundings.
2 They generate heat within their bodies.

Polar bears raise their fur to keep warm

Blubber is such a good insulator that seals lose very little heat to the surrounding water

Reducing heat loss

Animals can cut down the amount of heat that they lose to their surroundings but they cannot stop heat loss completely.

The skin plays a large part in reducing heat loss.

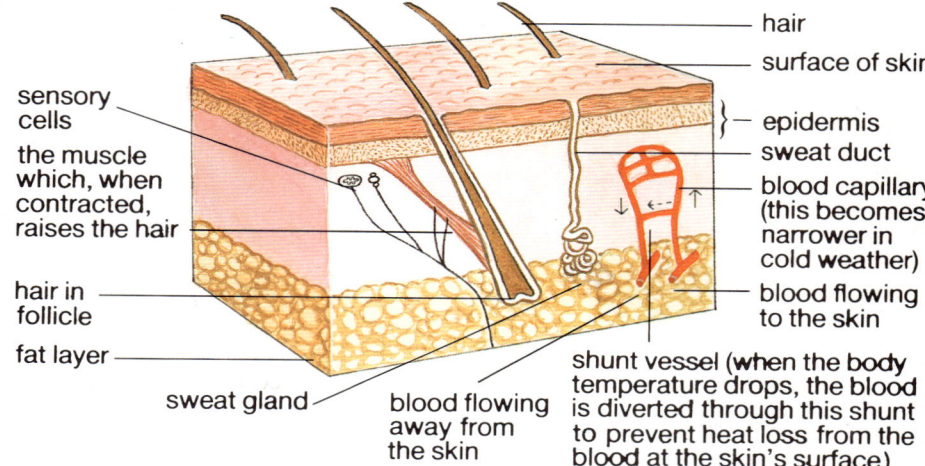

hair

surface of skin

sensory cells

the muscle which, when contracted, raises the hair

hair in follicle

fat layer

epidermis

sweat duct

blood capillary (this becomes narrower in cold weather)

blood flowing to the skin

sweat gland

blood flowing away from the skin

shunt vessel (when the body temperature drops, the blood is diverted through this shunt to prevent heat loss from the blood at the skin's surface)

Hairy or furry animals can raise and lower individual hairs by means of a special muscle near to the **hair follicle**. When the hairs are raised, more air can be trapped. Air is a poor conductor of heat and trapped air also cuts heat loss by convection. Radiation of heat is also reduced because the furry surface is cool. Feathers and clothes are good insulators for the same reasons. Wet fur is a poor insulator because the trapped air is replaced by water which is a much better conductor of heat.

If the blood vessels which run close to the surface of the skin become narrower, the amount of blood flowing through them is reduced. This is called **vaso-constriction** and it cuts down heat loss from the surface of the skin.

A third way that the skin controls heat loss is by having a layer of **fat** under the surface. Fat is a poor conductor of heat. All animals have some fat; those that live in a very cold environment have a much thicker layer called **blubber**.

Heat loss can be reduced by behavioural means, e.g. by going somewhere warm and sheltering. Humans shelter in buildings. Other animals, particularly small ones, dig burrows and line them with insulating material. Some animals huddle in groups. The air in the middle of the group is stiller and warmer than the surroundings.

Generating heat

One way to generate heat is by being more active. When muscles contract to produce movement they also produce heat. This heat is circulated round the body in the blood stream. **Shivering** is the constant contracting and relaxing of the muscles that the body sets off automatically as its temperature drops. Many animals also generate heat by increasing their basal metabolic rate (see p. 128).

If all these ways of keeping up core temperatures fail we will reach a dangerously low temperature and suffer from **hypothermia**. Hypothermia can be a danger to

- very elderly people who live in poorly heated houses. The risk can be made greater if they have too little to eat, have very little body fat, poor blood circulation and are not very active
- people who get stranded for long periods on mountains or down caves without adequate food and shelter
- small children—especially new-born babies. Their small size means that heat loss is quite rapid.

Keeping cool

Endotherms must also have ways of increasing heat loss if their core temperature rises. If a human's temperature rises to 42 °C or more death is almost certain. This is called **hyperthermia**.

Endotherms get too hot by being in a hot place. Hard exercise also increases our core temperature. During illness we often develop a fever as our bodies generate too much heat.

When humans get hot, they can remove some of their clothing. Animals flatten their fur and birds flatten their feathers so that their insulating layer of air is reduced and is less effective.

Some hairy animals like dogs do not have sweat glands on their skin. To lose heat they pant, and saliva on their tongue and the inside of the mouth evaporates which helps to cool them down

Going to a cool place with water is a good way of cooling down

The skin also regulates heat loss in the following ways.

1 Blood vessels near the surface of the skin widen to carry more blood. This is called **vaso-dilation**.

2 Sweat glands pour sweat onto the surface of the skin. As this evaporates heat is taken from the surface of the animal—cooling it down. Sweating does not work if the air is very humid (has a lot of water vapour in it).

 Take notes

Discuss in your group and write down a summary of the way animals control their temperatures.

 Project

Visit a sports shop and collect some information on clothing designed to keep mountain walkers warm.

Diffusion and osmosis

Diffusion

solid

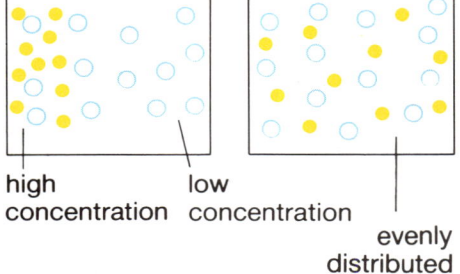

liquid

gas

Particles of matter are arranged differently in solids, liquids and gases.

Particles in solids are:

- packed closely together
- bonded to each other—but sometimes only weakly
- have vibrational energy
- are unable to move around or change places.

Particles in liquids are:

- packed loosely together
- weakly bonded to each other
- have more kinetic energy than in solids
- are able to change places with each other.

Particles in a gas:

- are separate from each other, with large distances in between
- move in all directions at very high velocity (speed)
- have the most kinetic energy
- are continuously colliding with each other.

The particles in liquids and gases are moving about all the time. As they do, they move from areas where there are a lot of them (**high concentration**) to areas where there are fewer particles (**low concentration**) until they are evenly distributed. This is known as **diffusion**.

high concentration   low concentration

evenly distributed

Here are some 'before and after' photographs of diffusion experiments.

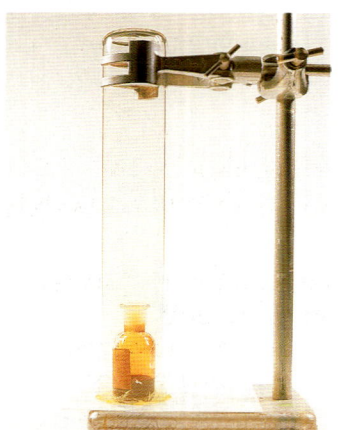

Bromine vapour from the liquid in the bottle has fully diffused after 30 minutes

A few drops of potassium permanganate added to the water have fully diffused after 20 minutes, without stirring

Exchange of gases

Gases like oxygen and carbon dioxide enter and leave our bodies through the **lungs**.

We **inspire** oxygen that is needed for respiration (see p. 139). Carbon dioxide is produced by the cells and is a toxic waste. Carbon dioxide must be **expired** (breathed out) from the lungs to prevent levels in the tissues from rising dangerously high.

The lungs are made up of thousands of small thin-walled air sacs called **alveoli**. They are covered by a dense network of **blood capillaries** (small, thin-walled blood vessels).

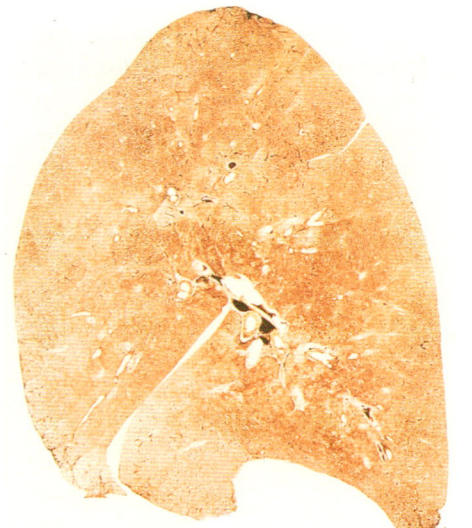

Section through a healthy human lung

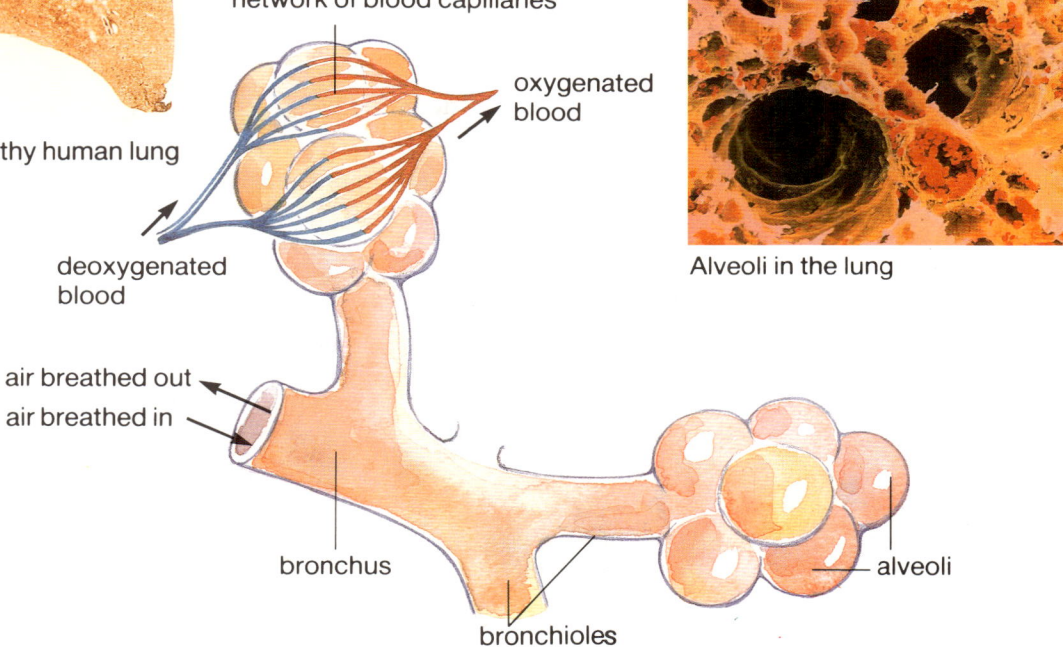

network of blood capillaries

oxygenated blood

deoxygenated blood

air breathed out

air breathed in

bronchus

bronchioles

alveoli

Alveoli in the lung

Carbon dioxide and oxygen pass between the lungs and the blood by diffusion.

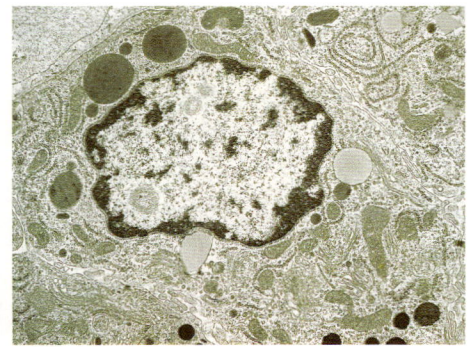

Electron micrograph of an epithelial cell. The large structure in the centre is the nucleus

Diffusion across membranes

Cells are the basic units of life. You should be familiar with the structure of a cell. (If you can't remember look it up in another book like *Understanding Science 1*.)

A cell is made up of structures, and the cytoplasm containing water and dissolved chemicals. Each cell is surrounded by a **cell membrane**. We want to investigate how water enters and leaves cells. In order to do this we use a simple model. Models are very useful in science as they copy a particular process and allow us to alter variables and learn more about the process. Visking tubing is a model of a cell as the tubing acts in some ways like the membrane of a living cell.

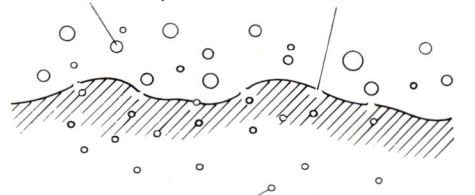

molecule too large to pass through membrane pore

pore in semi-permeable membrane

molecule sufficiently small to pass through membrane pore

Membranes are important in biology. They separate certain chemicals preventing unwanted chemical reactions and regulate the movement of other chemicals to allow reactions to be controlled more easily. Cell membranes are **permeable** to some molecules and **impermeable** to others. Small molecules, such as water molecules, can pass straight into the cell through the membrane pores, so the membrane is said to be permeable to these molecules. Larger molecules such as protein cannot pass through the membrane pores. The membrane is impermeable to these molecules. The membrane is said to be **partially permeable** as it allows some molecules to pass freely but acts as a barrier to others.

Collect

- 2 beakers (250 cm³)
- 3 pieces of Visking tubing
- protein and glucose solution
- distilled water

Can water cross Visking tubing?

1 Wet the Visking tubing under the tap to soften it.
2 Tie a knot in one end.
3 Fill the tube completely with the protein and glucose solution and tie the open end so that it is like a little sausage.
4 Put your cell model in distilled water.
5 Set up two more experiments as shown in the diagrams.
6 Leave them until the next lesson.

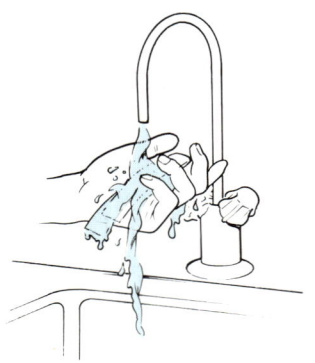

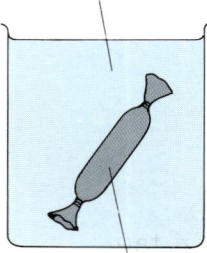

distilled water

Visking tubing full of protein and glucose solution

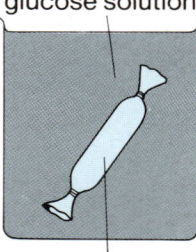

protein and glucose solution

Visking tubing full of distilled water

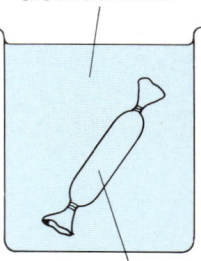

distilled water

Visking tubing full of distilled water

Q1 Write about your experiments. Include diagrams.

Q2 What do your results tell you about Visking tubing? (Hint: remember that not all molecules can pass freely across cell membranes.)

Q3 Can you explain your results based on your answer to Q2?

Q4 Do you think the model is a good one? Does it copy living cells?

Q5 Why did you use distilled water only in one of the experiments?

Osmosis

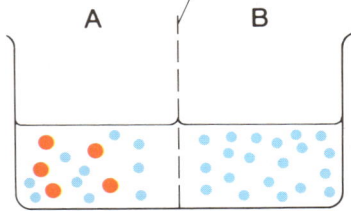

partially permeable membrane

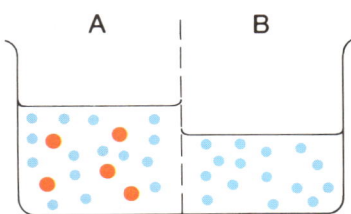

The name given to movement of water molecules across a partially permeable membrane from an area of high concentration to low concentration is **osmosis**.

Water molecules can pass across the membrane but large molecules such as sugar cannot. Water molecules move from side A to side B and from side B to side A.

As there are more water molecules on side B a larger number moves from side B to side A than from A to B.

This means that the total number of water molecules in side A rises and the amount of liquid in A goes up.

Osmosis in real cells

A potato is firm and wet when you cut into it. This is because a potato cell has a lot of water inside it which forces the cell contents to press against the cell wall. A cell in this stage is said to be **turgid**. The opposite to this is **flaccid**—the cell loses a lot of its water and the contents shrink away from the cell wall, leaving the potato soft.

◄ Red blood cell put in a solution which is more concentrated than the fluid in the cell. Water has passed out of the cell—the cell membrane wrinkles up

▼ Red blood cell put in a solution which is less concentrated than the fluid in the cell. Water has passed into the cell—and the cell membrane has stretched out

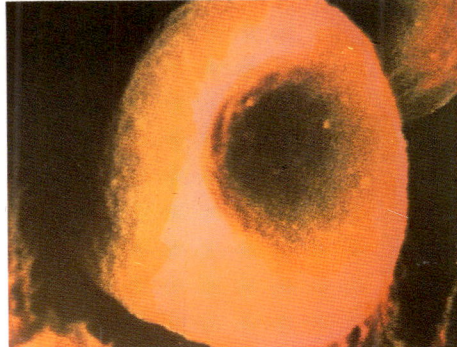

Collect

- potato
- test tube of concentrated salt solution
- test tube of distilled water
- graph paper

1 Cut two pieces of potato about 6 cm by 0.5 cm by 0.5 cm.
2 Measure their lengths accurately on millimetre graph paper. Weigh them.
3 Put one strip into each test tube and leave as long as possible.
4 Measure and weigh them again.

 *Take notes*

Write a report, including diagrams, on your experiment and explain your results; include words like osmosis, turgid, flaccid, partially permeable, impermeable in your description.

CHECKPOINT

Project

Find out about purification of sea water by osmosis.

Excretion and water balance

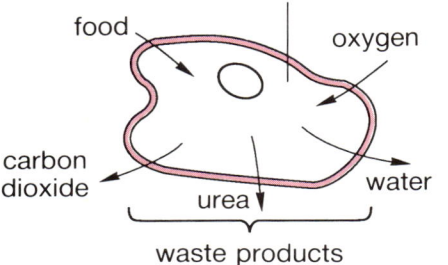

energy is produced for heat, movement, electrical activity of nerves, etc.

food

oxygen

carbon dioxide

water

urea

waste products

Unwanted chemicals

Metabolic reactions continually produce **waste products**. If these waste products are not eliminated they build up within the cells and disrupt the normal working of the cell, cause damage and eventually lead to death of the cell.

Excretion is the elimination of waste products from the body.

Urea and **carbon dioxide** are the main waste products that our bodies excrete. **Water** is also formed during some metabolic reactions. Excess water must also be excreted to keep the osmotic balance of the cell. Our bodies have four organs that carry out the job of excretion—they are called **excretory organs**. They are the **lungs**, **kidneys**, **liver** and **skin**.

The lungs

Collect

- glass tubing
- mouthpiece
- test tubes or small glass bottles
- limewater
- cobalt chloride paper
- thermometer
- safety glasses

1 Breathing tube

- Set up this apparatus using the equipment provided.

- Breathe gently in and out of the mouthpiece for a few seconds. (If you get a mouthful of limewater, don't worry! Rinse your mouth out and check that you have put the apparatus together properly.)

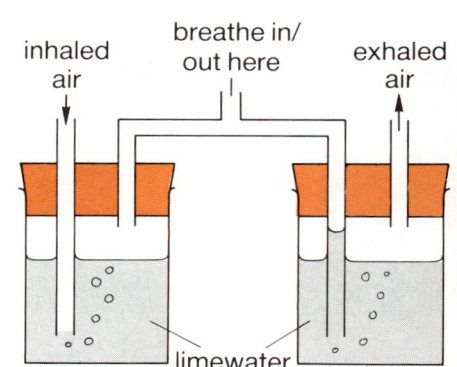

inhaled air

breathe in/out here

exhaled air

limewater

2 Cobalt chloride paper

- Breathe out onto a piece of cobalt chloride paper.
- Notice what happens.

3 Thermometer

- Note the temperature on the thermometer.
- Breathe onto the thermometer bulb and note the final temperature.

Q1 Make your own notes about these experiments.

Q2 What do the experiments tell you about air breathed in and air breathed out?

Respiration is an essential metabolic reaction that takes place in all our living cells. During respiration **energy** is obtained by breaking down digested food, which is a form of stored chemical energy. Much of our food is converted to glucose, which cells break down to carbon dioxide and water.

$$\text{glucose} \xrightarrow{\text{respiration}} \text{carbon dioxide and water and energy}$$
$$CO_2 \qquad + \qquad H_2O$$

We excrete the waste products **carbon dioxide** and **water** from our lungs (see p. 135).

Composition of air

	Inspired air	Expired air
Oxygen	21%	16%
Carbon dioxide	0.03%	4%
Nitrogen	79%	79%
Water vapour	variable	saturated with water

How is carbon dioxide transported out of the body?

Carbon dioxide (CO_2) is formed in the respiring tissues (e.g. muscle cells). It diffuses (see p. 134) into the red blood cells. An enzyme in red blood cells called carbonic anhydrase helps CO_2 to combine with water to form carbonic acid (H_2CO_3) which then splits up to form a hydrogen carbonate ion (HCO_3^-) and a hydrogen ion (H^+).

The hydrogen carbonate ion is converted back to carbon dioxide in the red blood cells as they pass through the lungs. The carbon dioxide then diffuses across the alveoli membrane and is expired from the lungs.

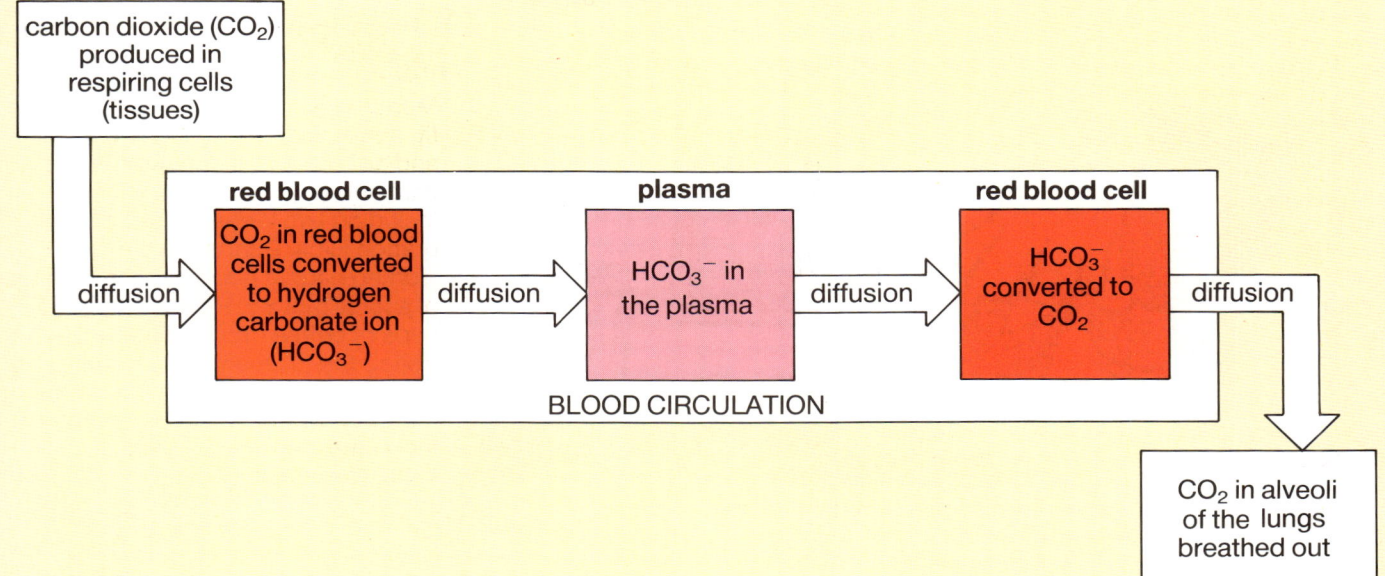

Discuss with your group and write down:

- the main features of the lungs
- what is meant by respiration.

The kidneys

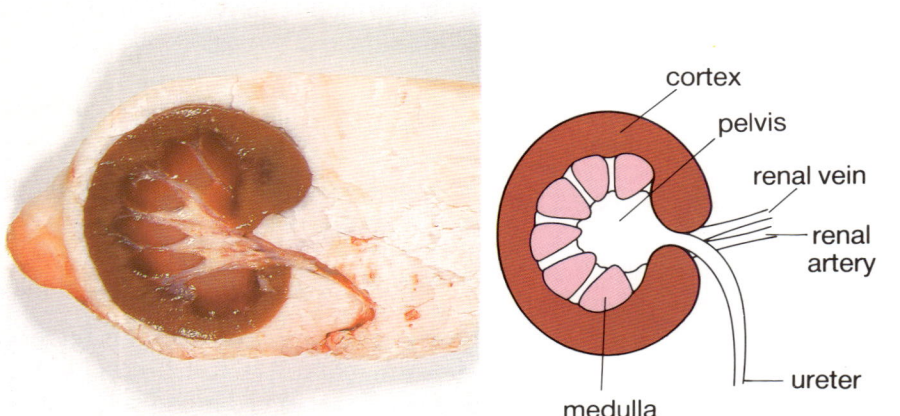

Collect
- lamb's kidney
- sharp knife

1. Slice the kidney carefully, lengthwise down the middle.
2. Compare the two cut surfaces with the photograph. Use the drawing and photograph to sort out different parts of the kidney.

cortex
pelvis
renal vein
renal artery
ureter
medulla

Q1 Draw a diagram of the kidney you have dissected. It might not look exactly like the one above. Label your diagram.

Water is lost from the body in **exhaled air**, in **sweat** and in **urine**. The regulation of water loss is carried out by the **kidneys**. The amount of **urine** produced depends on several factors. A normal amount of urine would be about 1500 cm^3–2000 cm^3 per day.

Think what an active young person might do during the day.

Murray got up at 7.00 am and being a fit, active type he jogged two miles and got rather hot and sweaty. He came home and had breakfast including two large glasses of fresh orange juice. During the day he worked hard at his job as the manager of a local factory. He drank several cups of coffee throughout the day, visited the toilet and ate normal meals. In the evening he met a few friends at his 'local' but got home fairly early and went to bed at 11.00 pm.

These are the amounts of water involved in various activities over a period of 24 hours.

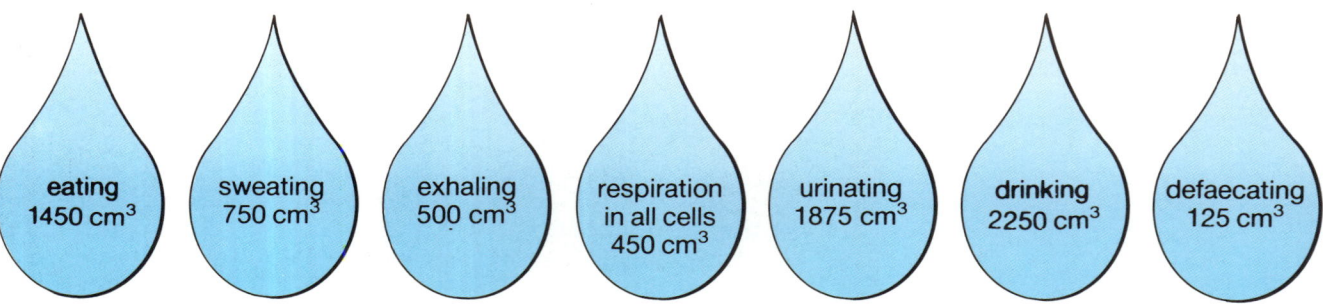

eating 1450 cm^3

sweating 750 cm^3

exhaling 500 cm^3

respiration in all cells 450 cm^3

urinating 1875 cm^3

drinking 2250 cm^3

defaecating 125 cm^3

Q1 Sort the 'uses of water' into a balance sheet of water gain and water loss. Check to see if gains and losses balance out.

Q2 Outline how the kidney works.

The liver

Urea and **bile** are formed in the **liver**.

amino acids from protein breakdown $\rightarrow$ mainly urea and other nitrogen-containing compounds

These toxic wastes are excreted from the body by the kidneys.

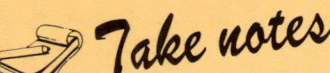

- resource sheet
- scissors
- glue

Where are the lungs, kidney and liver?

1 Cut out the lungs, kidney and liver and arrange them on the outline shape of a human figure.
2 Label the ureters, right and left kidneys, aorta, renal artery, vena cava, trachea, right bronchus, left bronchus, right lung, left lung.
3 Put the completed sheet in your book. Remember a heading.

Take notes

Discuss with your partner and write down:

- the main waste products excreted by the body
- how excretion takes place.

You could complete this table to summarise the information.

Organ	Excreted substance	How excretion takes place
liver	urea (excreted by kidney) bile (from breakdown of haemoglobin in blood)	defaecation and urination
skin		sweating
lungs		
kidneys		

Project

Find out how smoking affects the efficiency of the lungs.

Keep it steady

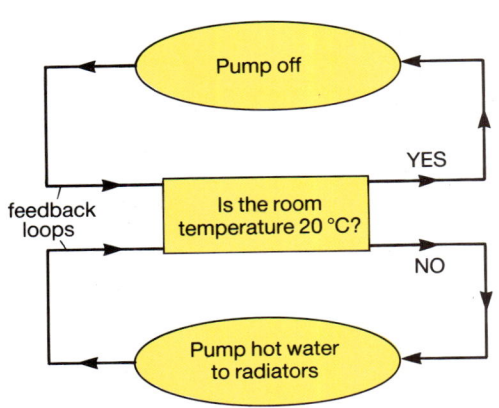

Start in the centre and follow either the NO loop or the YES loop. The thermostat is regularly checking the room temperature and electrical signals are sent out to the water pump either to start or stop pumping hot water to the radiators

Homeostasis

Homeostasis means keeping a steady state. It is the way the body regulates its internal conditions to keep all the cells functioning properly. We saw how temperature is regulated by homeostasis in Topic 7.2. Water balance (see Topic 7.4) is also regulated by homeostasis. Let us look at a simple example of homeostasis.

The central heating system in a house usually has a thermostat set to the 'normal' temperature. If the temperature falls below this then hot water is sent to the radiators until the room temperature returns to normal. Then the radiators are allowed to cool.

We can apply feedback loops to many different situations. In each case we have the same kind of mechanism.

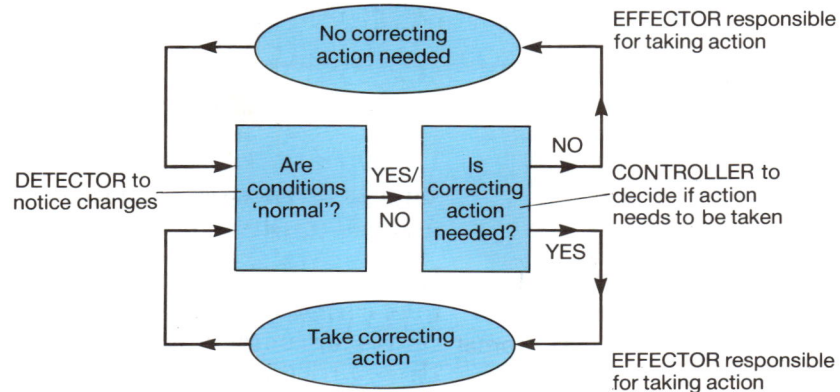

Controlling blood glucose

Glucose (sugar) is the basic 'fuel' for all the cells in the body. Even after a very sugary meal the amount of glucose in the blood does not rise very high. What keeps the amount of glucose constant? As it is not all used up as fuel for the cells, some of it is stored in the body. When the amount of glucose starts to rise (after a meal) the liver can store it until it is needed for cells in other parts of the body. The liver needs a special chemical messenger or **hormone** called **insulin** in order to store glucose. Insulin is produced by an organ called the **pancreas**. The liver converts glucose to **glycogen** (a kind of carbohydrate) as glycogen is more easily stored than glucose.

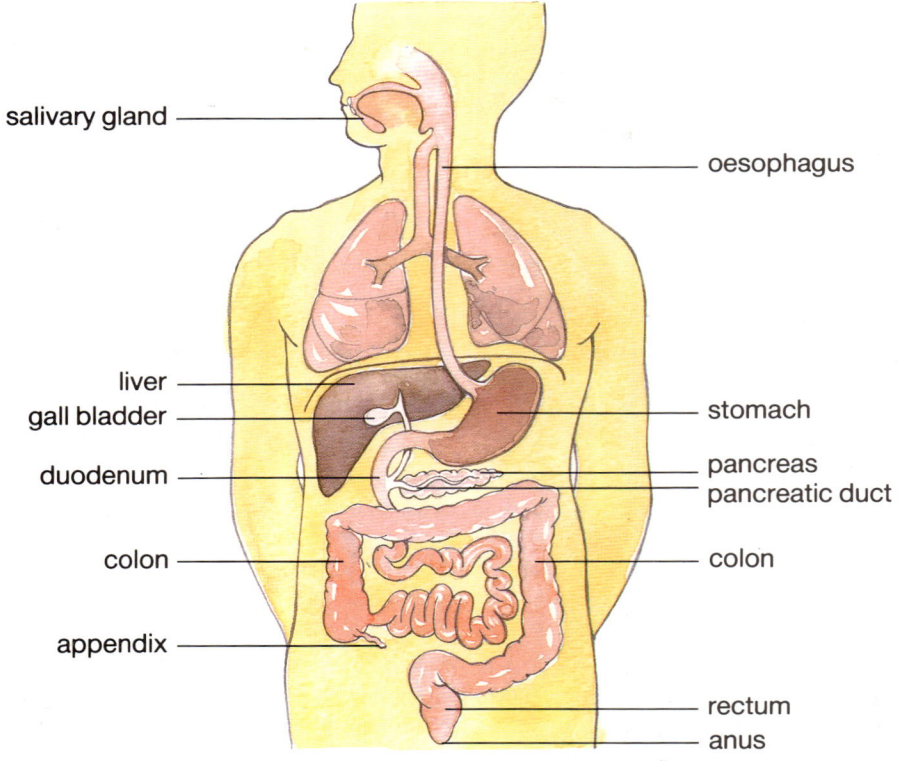

Thermoregulation (temperature regulation)
Make a flow-chart showing the feedback mechanism for thermoregulation. Follow the instructions on the sheet.

 **Take notes**

Discuss with your partner and write down:

- an everyday example of homeostasis (it need not be the central heating example) and an explanation of how it works
- the general mechanism for homeostasis
- a diagram showing how glucose is regulated.

Where is the temperature control centre?

How does the body know when to take correcting action? The centre for controlling **thermoregulation** is in a part of the brain called the **hypothalamus**. Nerve signals are received from various parts of the body in a certain part of the hypothalamus, giving information about the temperature of the blood.

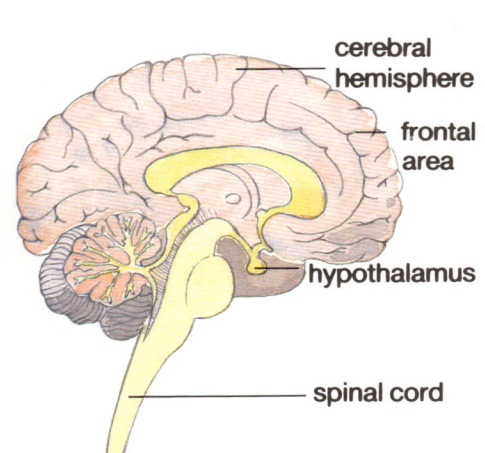

cerebral hemisphere

frontal area

hypothalamus

spinal cord

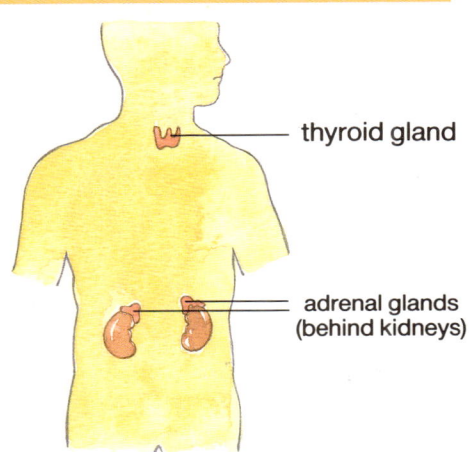

thyroid gland

adrenal glands (behind kidneys)

These glands send out hormones to the rest of the body

The hypothalamus is able to send signals to:

- various parts of the skin
- two different glands—the **adrenal glands** and the **thyroid gland**—which send chemical messages (hormones) to the rest of the body.

The skin alters the rate of temperature loss. The hormones stimulate the metabolic rate.

Can the 'thermostat' be altered?

During illness, we often develop 'a temperature'. What probably happens is that the hypothalamus resets the 'thermostat' to a higher value. We start to shiver to generate some more heat. Then we often sweat in order to cool down.

Other animals reset their hypothalamus more drastically. In winter, when food is short and the weather cold it is very difficult for some animals to maintain their body temperature. They solve this problem by going into a deep sleep called **hibernation**. During hibernation, body temperature can fall quite drastically and the metabolic rate drops. The hypothalamus readjusts itself to a lower setting. Some animals go even further and switch off their thermoregulation completely. This is a very risky business. The humming bird has trouble eating enough food to keep warm during the night, so it switches off its temperature controls and sleeps.

A dormouse hibernating

 **Take notes**

Discuss with your group and write down:

- the function of the hypothalamus
- the advantage of hibernation
- the basic 'mechanism' of hibernation.

When things go wrong

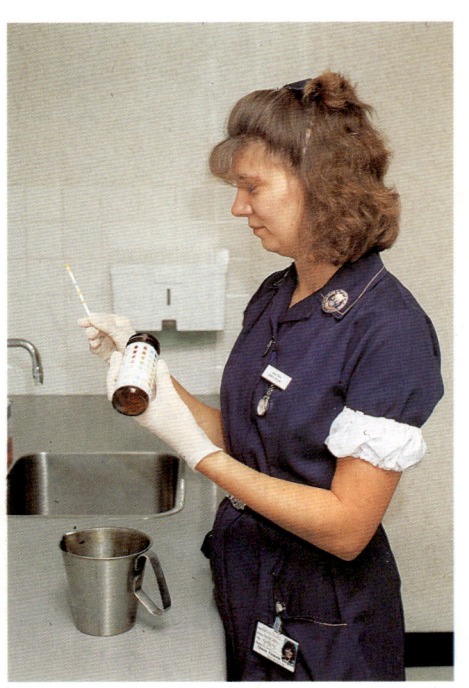

Sometimes the systems for regulating conditions go wrong. The composition of urine can often be a guide to changes going on in the body. Doctors use urine tests as part of the diagnosis of various illnesses. Damaged kidneys can result in blood and protein being released into the urine. **Diabetes** is a serious illness which develops when the body cannot store glucose properly (because of lack of insulin). Glucose passes into the urine.

Collect

- artificial urine samples
- biuret reagent
- Benedict's reagent
- test tubes
- tripod and gauze
- heat-proof mat
- Albustix (if available)
- safety glasses

Urine tests

1 Here are some tests and observations to make.

Testing for protein

1 Measure out 15 cm^3 of urine.
2 Add 2 cm^3 of biuret solution.
3 If colour changes from blue to mauve protein is present.

Testing for glucose

1 Measure 15 cm^3 of urine.
2 Add 2 cm^3 of Benedict's solution.
3 Place in boiling water for a minute.
4 Observe the colour. If it turns pink, a small amount of glucose is present. If it turns orange, or a green/brown solid appears, a large amount of glucose is present.

The appearance of urine

Check **colour**
—almost colourless
—pale yellow
—yellow
—amber
—other (describe)

Check **density**
(tilt the tube from side to side)
—is it slightly thick—does it cling to the side of the tube?

2 Test each of the samples of artificial urine for:
- protein
- glucose
- appearance.

If 'reagent strip' is available follow the instructions given.

3 Decide which sample came from each of the following cases:
A diabetic person
B healthy person who had drunk 1½ litres of water half an hour earlier
C normal healthy person doing nothing unusual
D person with kidney damage
E healthy person who has been doing hard physical work in hot weather.

Q1 Write a full report giving your results and conclusions.

 Project

Find out about the organ donor scheme and kidney transplants.

Problem

How does the elephant keep cool?

Elephants have a problem with thermoregulation. Consider the following facts:

1 Elephants are definitely big and heavy! Their surface area: mass ratio is small.

2 An elephant's internal temperature is 35.7 °C–36.7 °C but . . .

3 Elephants live in tropical areas such as Africa and India, where the summer temperatures can easily reach 48 °C and it gets very humid.

Q1 Explain why elephants have a problem losing heat.

Q2 How does an elephant lose heat?

Collect
- suitable apparatus

1 Think of a way that the elephant can cool down.
2 Devise an experiment to test your idea using the apparatus available.
3 Carry out your experiment once you have checked it with your teacher.

Q3 Write a report on your experiment. Did it work as you planned?

Q4 Explain how elephants keep cool in hot temperatures.

Talkabout

Regulate your life

All of these sports people are trying to regulate their internal conditions.

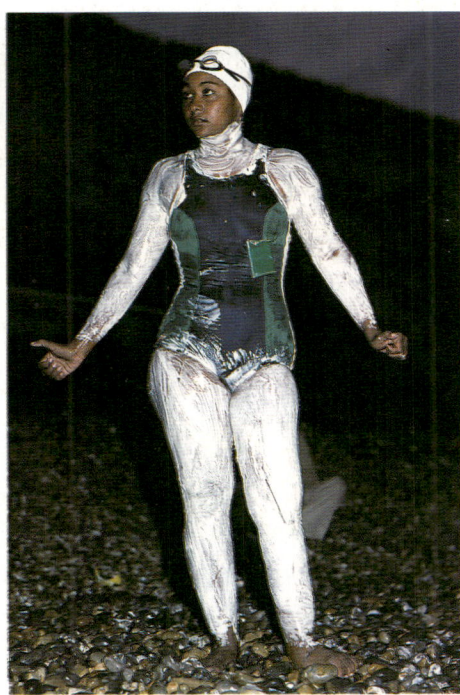

Decide which aspect they are trying to control and say how they are doing it.

1 The fur of arctic animals gives them very good insulation. A unit called a 'clo' is used to compare the insulation given by fur with the insulation given by human clothing. One 'clo' equals the insulation given by normal indoor clothing.

The graph below shows the insulation given by the fur of various animals in 'clo' units.

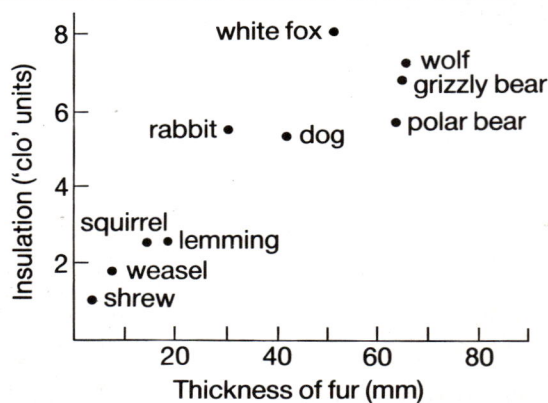

a Which animal has the thickest fur? [1]
b Which animal has least insulation? [1]
c The maximum insulation which is provided by arctic survival clothing is 6 'clo' units.
  i Which animals can survive colder conditions than a human dressed in such clothing? [1]
  ii Suggest **one** reason why it is not practical to provide arctic survival clothing with a 'clo' value greater than 6 units. [1]
d The diagram shows the temperature at various points inside an arctic dog's leg.

Suggest why it is an advantage to the dog to have such a low temperature in its feet. [1]
NEA 1989 Biology

35 °C
• 14 °C
8 °C
0 °C

2 The following diagram shows an experiment done by some students in a school.

Hot water is poured into the beaker and the temperature of the water is taken every minute for 10 minutes. The results are shown in the table.

The experiment is repeated with different types of material around the beaker.

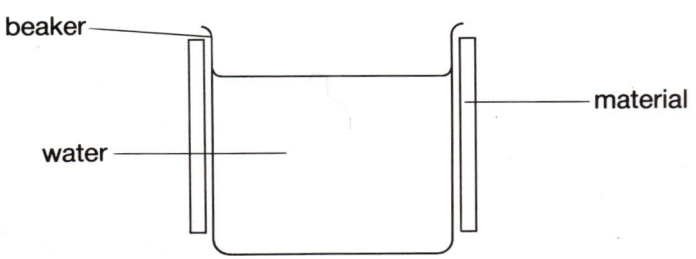

beaker
material
water

Time in minutes	Temperature in °C when material around beaker is			
	no material	wool	glass fibre	cotton
0	95	95	95	95
1	71	85	84	82
2	52	76	74	71
3	37	68	65	62
4	30	61	57	54
5	25	55	60	47
6	23	50	44	41
7	23	45	39	36
8	23	41	35	32
9	23	37	32	29

a What was the temperature of the water after 6 minutes when cotton was used? [1]
b What was the *drop* in the temperature of the water during the experiment when glass fibre was used? [1]
c Name **two** important pieces of equipment *not* shown in the diagram that are needed for this experiment. [2]
d What effect, if any, would there be on the *drop* in temperature in the first minute if
  i *hotter* water had been used? [1]
  ii *more* water of the *same temperature* had been used? [1]
e In which beaker had the water stopped cooling down *before* the 9 minutes was up? [1]
f i What was the temperature of the room where the experiment was done? [1]
  ii Explain your answer. [1]
g i **One** of the readings taken when *glass fibre* was being used is a mistake. Which one is it? [1]
  ii *Plot a graph* of the results for *glass fibre*. Draw the cooling curve for the results. [5]
  iii What have you done about the mistaken reading? [1]
MEG 1988 Specimen Extension Paper Science

8

Start small

What is an atom?

A lot of space

All substances are made up of **atoms**. Sometimes there is just one **element** containing the same kind of atoms and sometimes the elements join together to form **compounds**. You will remember how elements can be arranged into a Periodic Table.

An atom is the smallest particle of an element that can exist on its own.

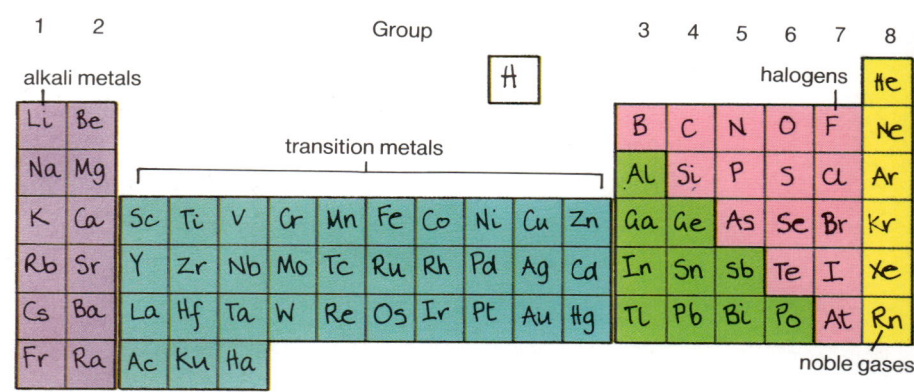

Each artifact contains billions of identical atoms of iron, gold or silver

An atom of gold is different from an atom of silver. Atoms are so small that scientists need to study the properties of the substances so that they can develop a model of the structure of atoms. A hundred years ago many people did not believe that matter was made of atoms. Even today we are still trying to find out more about the structure of atoms.

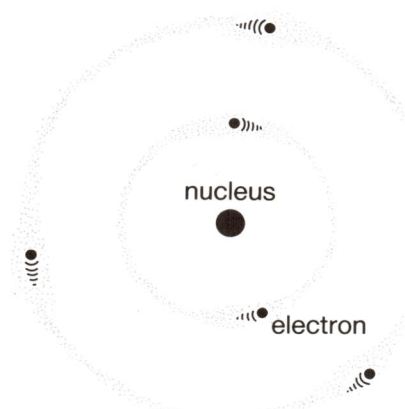
nucleus

electron

The structure of atoms

An atom is mainly empty space. Most of its mass is concentrated in its centre in a region called the **nucleus**. The nucleus has a positive charge which is balanced by negatively charged **electrons**. We can think of electrons *orbiting* the nucleus rather like planets round the sun.

An atom as a whole is electrically **neutral**. The positive charge of the nucleus is equal to the negative charge of the electrons.

Each electron has a negative charge of -1. If there are five electrons the total charge will be -5. This also tells us that the nucleus of this atom must have a charge of $+5$.

Gold and silver atoms compared

The differences between elements are due to variations in the positive charge of the nucleus and the number of orbiting electrons needed to balance it, and the size of the atoms.

In the Periodic Table each element is represented by two numbers.

The smaller number is the **atomic number**. It tells us the size of the positive charge on the nucleus. The same number tells us the **number of electrons** orbiting the nucleus, keeping the atom neutral.

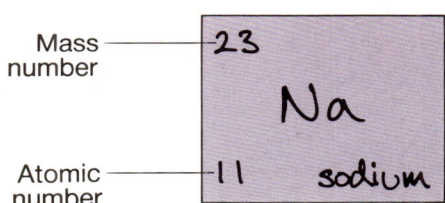

Mass number —— 23

Na

Atomic number —— 11   sodium

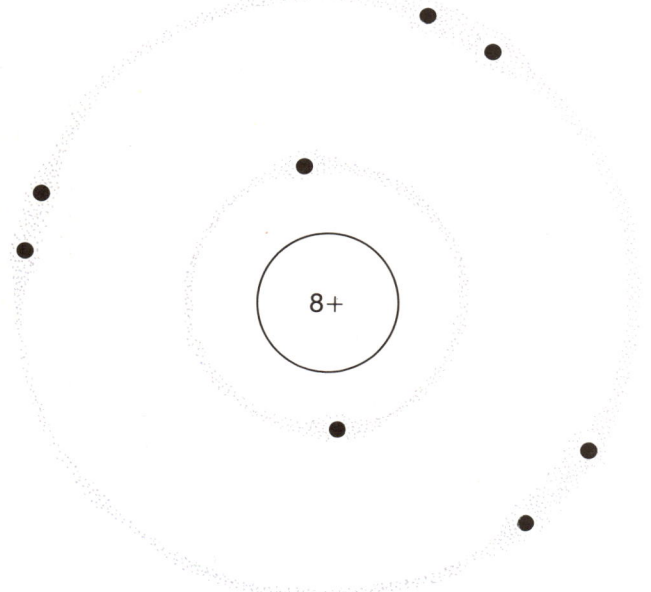

Oxygen has an atomic number of 8. This means that the nucleus has a charge of 8+. To remain neutral there must be 8 electrons in orbit around the nucleus

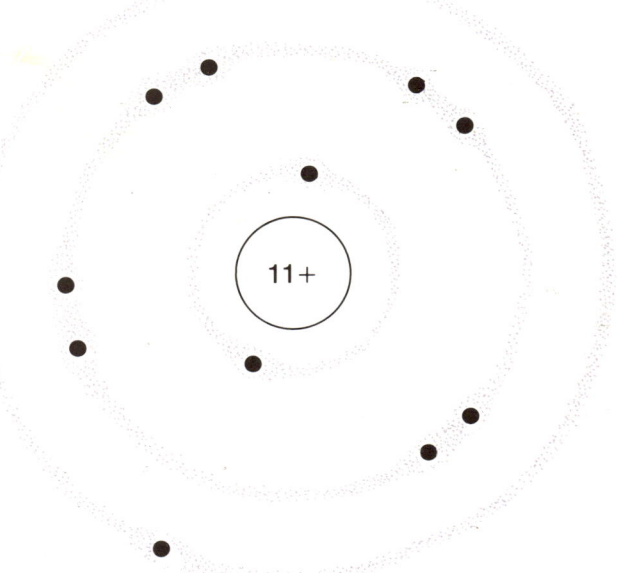

Sodium has an atomic number of 11. This means that the nucleus has a charge of 11+. To remain neutral there must be 11 electrons in orbit around the nucleus

An atom can gain or lose electrons. It is then no longer neutral and it is described as an **anion** (negative ion) or a **cation** (positive ion).

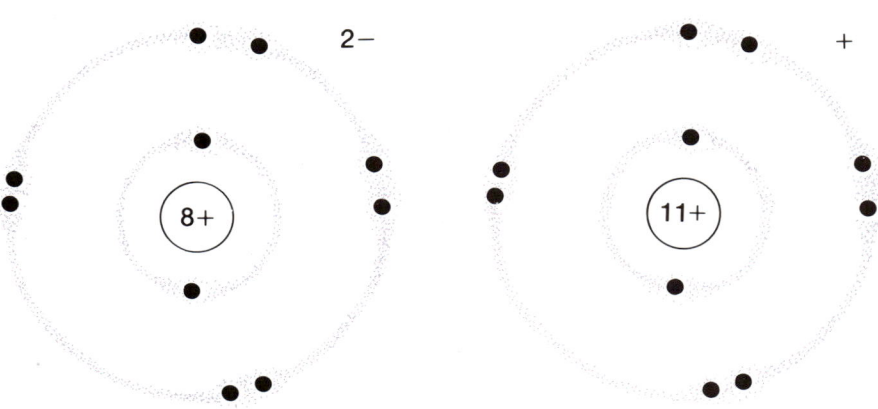

Oxygen can gain two electrons. It becomes the ion O^{2-}. It has 10 negative charges but only 8 positive charges

Sodium can lose one electron. It becomes the ion Na^+. It has 10 negative charges but 11 positive charges

Electron arrangement

Electrons are not arranged randomly around the nucleus. They occupy different orbits or **shells**. The term used to describe this is the **electron configuration**.

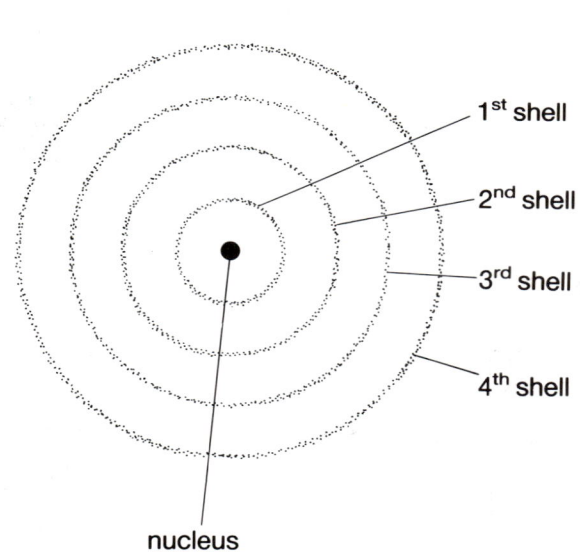

1st shell
2nd shell
3rd shell
4th shell

nucleus

This model shows how electrons are thought to orbit at different distances from the nucleus

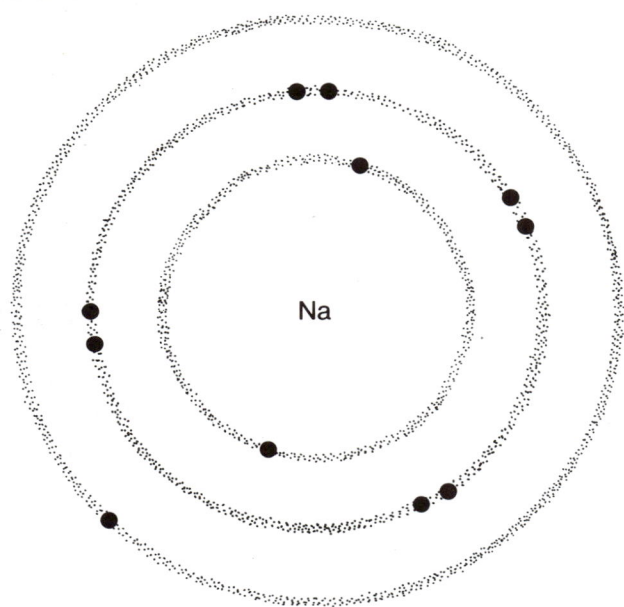

Na

The sodium atom has 11 electrons: 2 in the first shell, 8 in the second shell and 1 in the third shell. This is written as 2,8,1 and is known as the electron configuration

The innermost shell for any of the first 20 elements of the Periodic Table can hold a maximum of **two** electrons.

The outer shells can hold a maximum of **eight** electrons.

1 Copy the table below.
2 Use the Periodic Table to fill in your table.

Atomic no.	Element	Charge on nucleus	No. of electrons
1			
6			
	Gold Au		
	Silver Ag		
		29+	
		9+	
			35
			14

 **Take notes**

Discuss with your partner and write down details on:

● an atom
● a nucleus
● an electron
● how to work out the electron configuration of an atom.

Drawing atoms

● Write down the electron configurations of hydrogen, fluorine, magnesium and argon atoms.

● Draw diagrams to show the electron arrangements in the above atoms.

 **Project**

Find out about the plum pudding atom and how it was squashed by Rutherford.

Weighing atoms

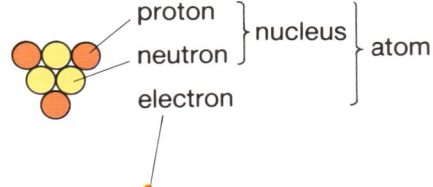

proton
neutron } nucleus } atom
electron

Into the nucleus

Atoms are mainly empty space, and their mass is very small. The heavy part of an atom is the **nucleus**. During this century scientists have been investigating the nucleus to see if it is made of smaller particles. In the 1930s it was shown that the nucleus was made up of **protons** and **neutrons**. Protons have a positive charge. Neutrons carry no charge; they are neutral.

Since then scientists have been trying to find out if any other particles are produced when an atom is split. They have identified over 200 to date but it is likely that these all form three basic groups of particles. These are known as **hadrons** (made up of **quarks**), **leptons** and **photons**. Scientists are still investigating the structure of the atom but we can learn a lot about the way atoms and molecules behave if we concentrate on neutrons, protons and electrons.

The tunnel of the proton accelerator at CERN, Switzerland. Very strong electric and magnetic fields make the protons travel round a ring at very high speed. The scientists then investigate the particles produced by their collisions

Atoms and their sub-atomic particles are so small that it is difficult to give their mass in grams. Instead it is easier if we call the mass of a proton one atomic mass unit (a.m.u.). All other atomic masses are compared to the mass of one proton. A neutron has the same mass as a proton but electrons are nearly 2000 times lighter (so small their mass is often ignored).

Particle name	Mass (a.m.u.)	Charge	Region
proton	1	1+	nucleus
neutron	1	0	nucleus
electron	$\frac{1}{1836}$	1−	shell

We saw (on page 151) how the Periodic Table gives two numbers for each element. The atomic number gives the positive charge of the nucleus. Protons carry this charge. The atomic number is therefore the **number of protons**. The larger number is the **mass number** which is the **total number of protons and neutrons** in a nucleus. You can use the two numbers to work out how many protons, neutrons and electrons there are in an atom of any element. (*Remember* the number of protons and electrons must be the same as the atom is electrically neutral.)

Let's work out the number of **neutrons** in a sodium atom.

Number of protons   = atomic number = 11
Number of neutrons = mass number − number of protons
$$= 23 - 11$$
$$= \textbf{12 neutrons}$$

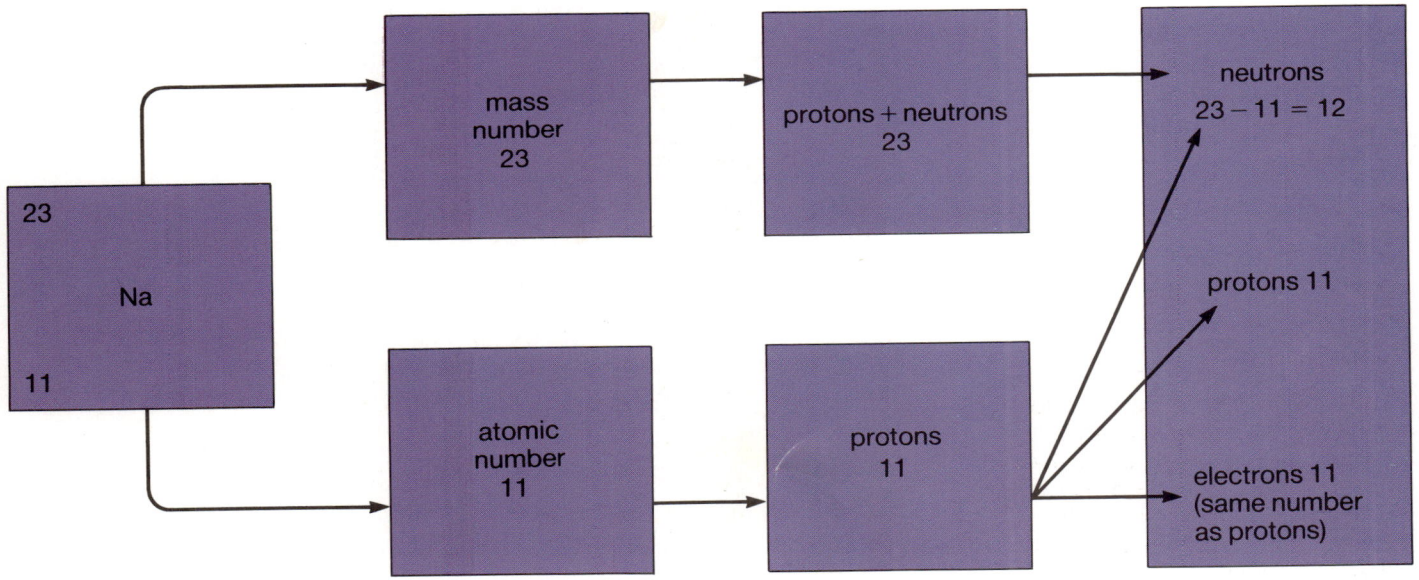

Q1 Use the Periodic Table to work out the number of protons, neutrons and electrons for the first 20 elements (hydrogen to calcium).

Q2 What do you notice about the electron shells of the noble gases (group 8)?

✍ *Take notes*

Discuss with a partner and write down:

- what particles the nucleus contains
- the mass and charge of a proton, a neutron and an electron
- how to work out the numbers of particles in an atom.

Isotopes

The photograph shows a gas jar of the element chlorine. As it is an element, all the gas molecules are made up of chlorine atoms. Although the gas *looks* the same throughout the jar, the chlorine atoms making up the gas are not exactly the same.

There are two types of chlorine atom:

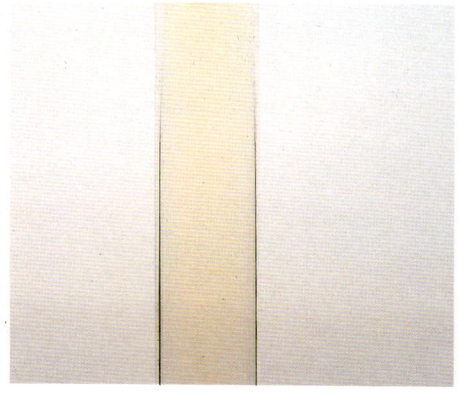

We can construct a table to show the differences between the two chlorine atoms.

Atom	No. of protons	No. of neutrons	No. of electrons
$^{35}_{17}$Cl	17	18	17
$^{37}_{17}$Cl	17	20	17

Both chlorine atoms have the same number of protons and electrons. The difference lies in the number of neutrons. Atoms which have the same atomic number but different mass numbers (i.e. a different number of neutrons) are called **isotopes**. As isotopes have the same atomic number they have similar chemical properties. This is why the chlorine in the gas jar looks to be the same. Chlorine isotopes are **stable** (unreactive). Some isotopes are very **unstable**. Many are **radioactive**. (You should remember what this means—if not look up in a book like *Understanding Science 3*.) Scientists have found a use for these isotopes in bomb-making, medicine, carbon-dating and many other ways.

Hydrogen has three isotopes

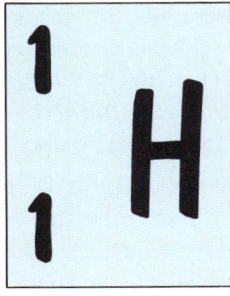 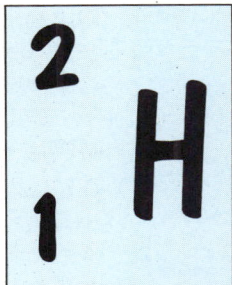 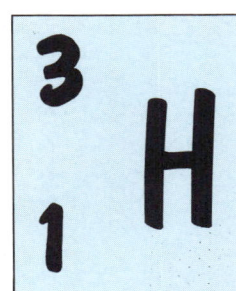

Q1 Construct a table like that for chlorine to show the number of particles in each of the hydrogen isotopes.

Q2 Draw electron configurations for the three hydrogen isotopes.

Use reference books to help you answer the following questions.

Q3 Why was deuterium ($_1^2$H) of great importance during the Second World War?

Q4 Why is $_6^{14}$C of use to archaeologists?

Q5 Why is $_{53}^{131}$I of use in medicine?

 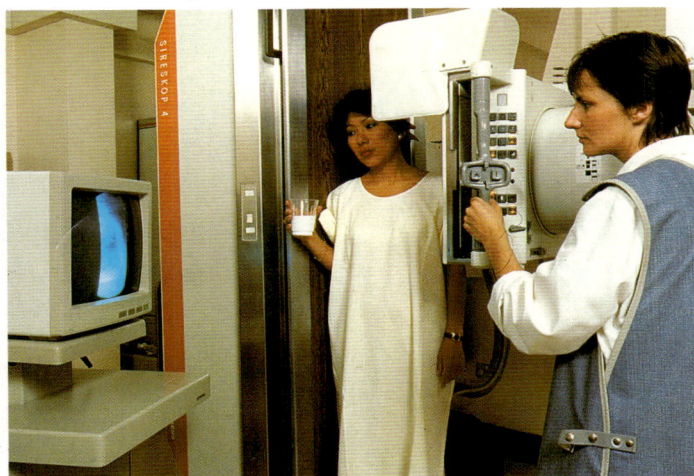

Q6 Write an article about one of these uses of isotopes.

 Project

Find out about an electron microscope. How does it work? What is it used for? What advantages does it have compared to a bench microscope?

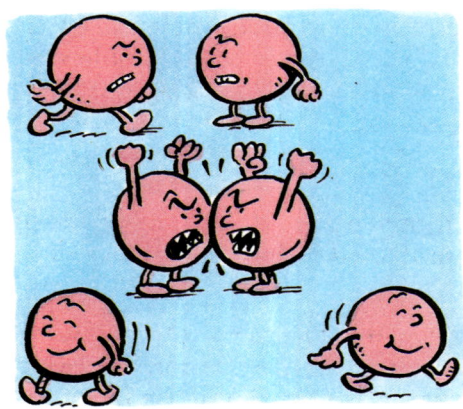

These two argon atoms have collided and moved apart

Simple molecules

The noble gases are very unreactive (**stable**) elements. This is because their atoms have full outer shells of electrons. If two atoms of a noble gas collide they bounce apart and continue to move in different paths. It seems that other elements react so that their atoms will achieve full outer shells like those of noble gases.

What happens if two hydrogen atoms approach each other? A hydrogen atom has only one electron in its outer shell and so needs one more electron to fill its shell. As the hydrogen atoms get close to each other both nuclei attract the other's electron. The two atoms end up **sharing electrons**.

The shared electrons form a **covalent bond** which bonds the two hydrogen atoms together to form a hydrogen **molecule**. Covalent bonds are very strong. The molecule contains two hydrogen atoms so the **molecular formula** is H_2.

A hydrogen molecule can be represented as H—H. The line drawn between the two atoms represents the covalent bond. This model of the hydrogen molecule helps us to explain its chemical properties.

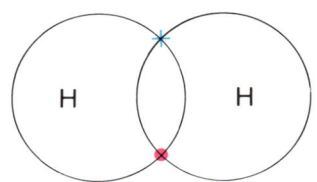

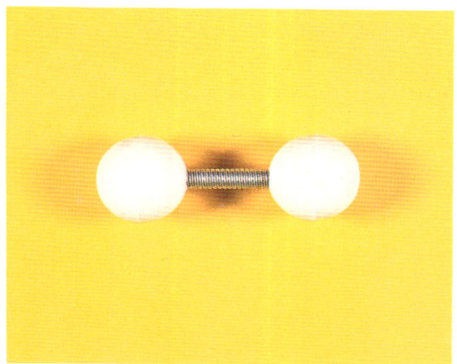

A hydrogen molecule

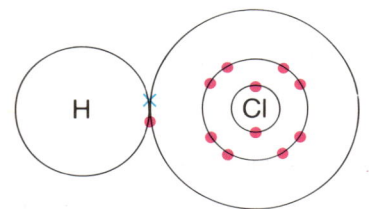

A hydrogen chloride molecule

Covalent bonds can also be formed between atoms of **different elements**, for example, hydrogen and chlorine. The result is a molecule of **hydrogen chloride**.

There are **three atoms** in a molecule of water. The electron diagram of an oxygen atom shows that it has six electrons in its outer shell. It therefore needs to gain two to be full. If oxygen molecules collide with hydrogen molecules then two hydrogen atoms will be needed to share electrons with the oxygen atom. **Two covalent bonds** are formed.

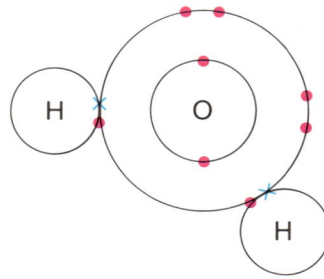

A water molecule

The oxygen atom now has eight electrons in its outer shell and each hydrogen atom has two electrons. The three atoms have a noble gas structure. The molecular formula for water is H_2O because the molecule contains two hydrogen atoms and one oxygen atom.

Multiple covalent bonds

Imagine oxygen atoms colliding with each other. Each atom needs to gain two electrons. This can be accomplished if they share **four electrons**, two from each atom.

Instead of one covalent bond being formed between the atoms there are two. This is called a **double bond**. Each shared pair of electrons forms a bond. The molecule can be drawn as $O = O$ and the formula for the oxygen molecules is O_2.

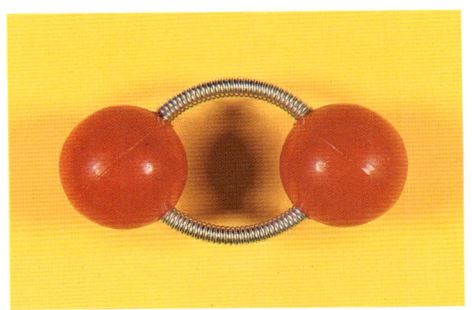

An oxygen molecule

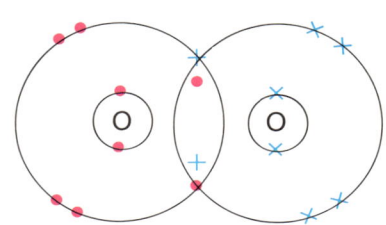

Collect
- model-making kit

Making molecules

1 Use the kit to make models of the molecules listed:

- methane (CH_4)
- ammonia (NH_3)
- oxygen (O_2)
- nitrogen (N_2)
- carbon dioxide (CO_2)
- propane (C_3H_8)

(Watch out for the triple bond!)

Q1 Draw diagrams to show how covalent bonds are formed between the following atoms and write the molecular formula for each:
  a fluorine
  b ammonia (N and H atoms)
  c methane (C and H atoms).

 *Take notes*

Discuss with your partner and write down:

- what happens when two hydrogen atoms combine
- what happens when hydrogen and chlorine atoms combine
- what a covalent bond is
- three other covalent molecules, with drawings
- what double and triple bonds are.

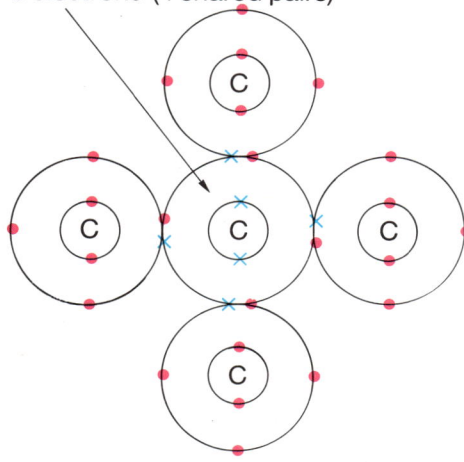

the central carbon atom has 8 electrons (4 shared pairs)

the outer carbon atoms have 5 electrons (1 shared pair)

Carbon atoms bond to form the macromolecule diamond

Giant covalent structures

Some covalent substances exist as giant structures called **macromolecules**. (*Macro* is from the Greek *Makros* meaning great.) These macromolecules are made up of a **giant network** of covalently bonded atoms.

Diamond is a macromolecule. It is an extremely hard substance—in fact it is one of the hardest substances known! It is made only of carbon atoms. Each carbon atom has four electrons in the outer shell so needs to gain four to have a full outer shell. Each carbon atom forms covalent bonds with four other carbon atoms.

The central carbon atom now has the eight electrons in its outer shell but all the other carbon atoms still need three more electrons to achieve this. Each atom bonds with three more carbon atoms. This bonding pattern repeats to form a diamond macromolecule.

Diamonds have uses in addition to their considerable value as jewels. It is only the colourless or faintly coloured, transparent diamond that is used to make jewellery. The coloured forms are used for rock drills, for lathe tools and, when powdered, for polishing. Diamond is a very hard and strong substance.

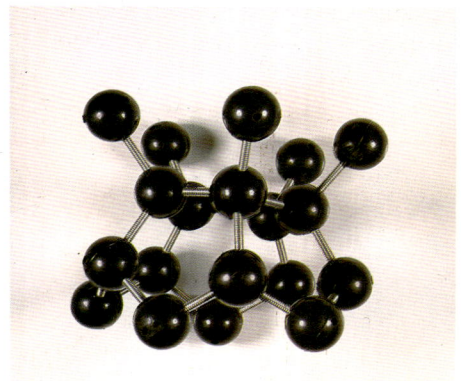

A diamond macromolecule

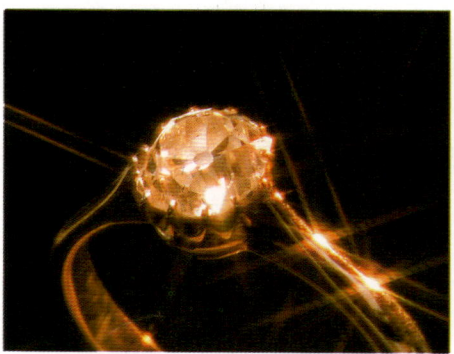

a   Diamond ring

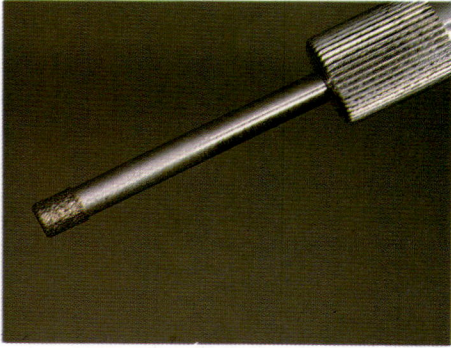

b   Cutting tool made from diamond

Graphite is another macromolecule that is made only of carbon atoms. It is a soft substance. It is used as a lubricant, in pencil leads, in electrical components and in paints. It is very different to diamond which is surprising because they are both carbon macromolecules.

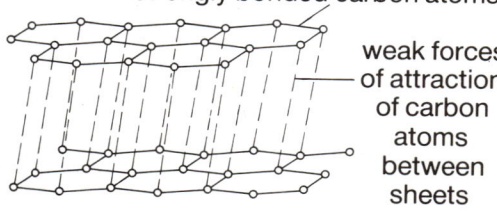

strongly bonded carbon atoms

weak forces of attraction of carbon atoms between sheets

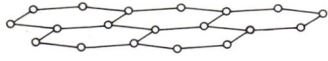

Graphite

a   Graphite cores in a brush connector used in aircraft generators

b   Graphite pencil leads

Look at the molecular structure of diamond and graphite. Can you suggest why diamond is so hard and strong whereas graphite is soft and brittle? It is because the graphite macromolecule is made up of many 'sheets' of carbon atoms. The covalent bonding between the carbon atoms in each sheet is strong but only very weak attraction exists between the sheets. One sheet can move relatively freely over another. This gives graphite its unusual physical properties.

Allotropes

When an element exists in two or more crystalline forms the different forms are called allotropes.

Collect

- samples of graphite
- test tube
- test tube holder
- test tube rack
- bunsen burner
- heat-proof mat
- circuit to test electrical conductivity
- cube of graphite
- hammer
- safety glasses
- set of cards

1 Investigate the following properties of graphite:
   - melting point
   - electrical conductivity
   - density
   - hardness.
2 Using what you have read earlier in the unit and a science reference book find out as much about the properties of diamond as you can.
3 Look at the information on the cards concerning diamond and graphite. Match each statement to the correct reason and the correct form of carbon. Write your answers in the form of a table. Use the headings *Property* and *Explanation*.

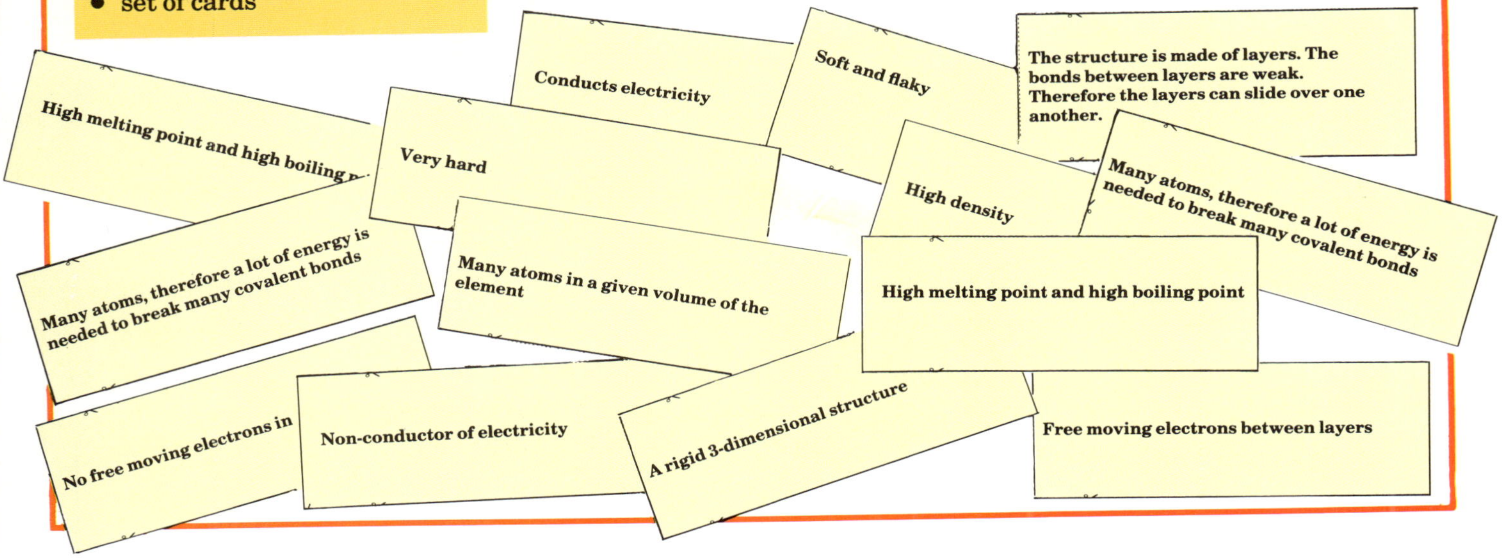

Conducts electricity

Soft and flaky

The structure is made of layers. The bonds between layers are weak. Therefore the layers can slide over one another.

High melting point and high boiling point

Very hard

High density

Many atoms, therefore a lot of energy is needed to break many covalent bonds

Many atoms, therefore a lot of energy is needed to break many covalent bonds

Many atoms in a given volume of the element

High melting point and high boiling point

No free moving electrons in

Non-conductor of electricity

A rigid 3-dimensional structure

Free moving electrons between layers

Q1 What are the names of the allotropes of carbon?

Q2 For each of the photographs opposite explain in terms of their structure why diamond or graphite is used.

Q3 Find out about the allotropes of sulphur. Find out what they are called and what their properties are? How are the different allotropes made? Your teacher may show you how to make these allotropes.

Project

Find out the chemical name for sand and how the atoms are bonded together. (Sand is a giant covalent structure.)

Ionic compounds

The element sodium

- **Losing electrons from an atom forms a cation (positive ion)**
- **Gaining electrons forms an anion (negative ion)**
- **The number on the charge is the same as the number of electrons lost or gained**

Give and take

Atoms may either gain or lose electrons in order to achieve full outer shells.

Sodium (Na) has an electron configuration of 2,8,1. If a sodium atom gained seven electrons *or* lost one electron it would have a full outer shell. It takes less energy to lose one electron than to gain seven. The electron configuration is then 2,8.

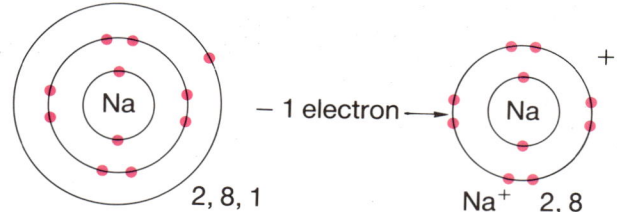

We have seen that atoms are electrically neutral because they have equal numbers of protons and electrons. A sodium atom has 11 protons and 11 electrons. If the atom loses one electron it is no longer neutral but has an overall charge of +1 because it now has 11 positive protons and only ten negative electrons. The particle is now called a sodium **ion** and is written Na^+. This positive ion is called a **cation**.

Chlorine has an electron configuration of 2,8,7. It takes less energy for the chlorine to gain one electron in its outer shell than to lose seven. The chlorine atom becomes a chloride **ion** when it gains an electron because it now has a total of 17 protons and 18 electrons. It has the formula Cl^-. This negative ion is called an **anion**.

Not all atoms form ions with only one charge. Sulphur has the electron configuration 2,8,6 and gains two electrons giving an overall charge of -2. There are 16 protons and now 18 electrons forming a sulphide anion S^{2-}.

Copy and complete the following.

Atom	Electron configuration of atom	Electron configuration of ion	Formula of ion
Al Ca Li O F N			

Ionic bonding

What happens to the lost electron when a sodium atom becomes a sodium ion? Where does a chlorine atom get an electron to form a chloride ion?

Imagine a sodium atom colliding with a chlorine atom.

Na 2,8,1

1 electron transfer

Cl 2,8,7

As the sodium atom needs to lose one electron and the chlorine atom needs to gain one electron, the sodium 'gives' its electron to the chlorine atom. The two atoms become ions which have opposite charges

The oppositely charged ions attract each other and form a giant **ionic lattice** or **network** of sodium and chloride ions which make up sodium chloride **crystals**.

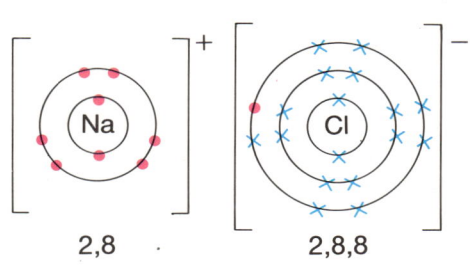

$\begin{bmatrix} Na \\ 2,8 \end{bmatrix}^+ \quad \begin{bmatrix} Cl \\ 2,8,8 \end{bmatrix}^-$

Electronic configuration of sodium chloride

A model showing the lattice structure of sodium chloride

Salt (sodium chloride crystals)

Each positive ion is surrounded by negative ions and vice versa. The forces of attraction between these two ions are very strong. As a result sodium chloride has a high melting and boiling point.

The formula for sodium chloride is **NaCl** because only one atom of sodium is needed to give its electron to each chlorine atom.

This is not always the case. When lithium oxide is formed each lithium atom needs to lose one electron but the oxygen atom needs to gain two electrons.

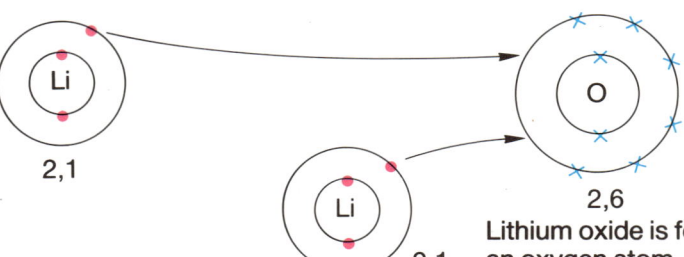

Lithium oxide is formed when two lithium atoms donate one electron each to an oxygen atom

This means that two lithium atoms are needed for every oxygen atom. The formula for lithium oxide is **Li₂O**. The ions formed are $2Li^+$ and O^{2-}.

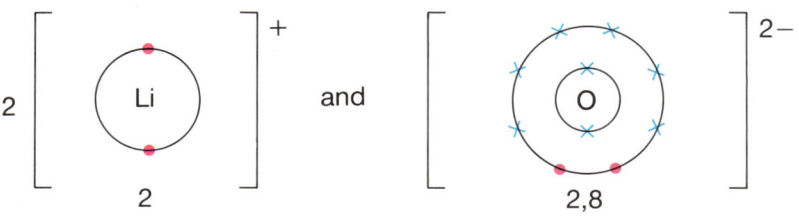

The overall number of charges remains the same: two positive charges from two lithium ions and two negative charges from one oxygen ion. The lithium chloride is an electrically neutral ionic compound and is stable.

Q1 Draw bonding diagrams to show the electron transfer and the ions produced when the following compounds are formed:

    a lithium chloride
    b sodium oxide
    c magnesium oxide
    d aluminium chloride
    e potassium sulphide.

Q2 Write the formula for each compound.

Discuss with your partner then:

- write down what an ion is
- describe how sodium and chlorine change their electron configuration to obtain a noble gas structure
- make a drawing of a sodium and a chlorine atom transferring an electron
- write down what an ionic bond is.

Metallic bonding

Metals also consist of giant lattices of ions. In potassium the metal ions pack closely together in the lattice. Potassium has the configuration 2,8,8,1 so each atom can lose one electron. This means that there are a large number of released electrons in the lattice. These electrons are free to move about between the positive ions that are formed. The electrons are **delocalised** and belong to all the positively charged metal ions.

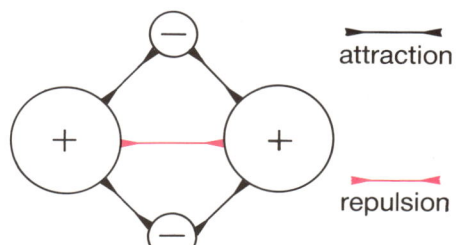

freely moving electrons

The delocalised bonding in metals

attraction

repulsion

Although positive ions repel one another the lattice is held together by the stronger forces of attraction between the negatively charged free electrons and the positively charged metal ions

Q1 Discuss with your partner what you know about the properties of metals (you can use other reference books to help you).

Q2 Explain the properties of metals in terms of their structure and bonding.

Growing and observing crystals

Collect

- microscope
- slides
- glass rod
- solutions
- test tubes and rack
- lead nitrate or lead ethanoate solution
- zinc foil
- silver nitrate solution
- copper foil
- safety glasses

Ionic crystals

1 Place a drop of saturated solution on a slide.
2 Observe through a microscope.
3 Draw the shapes of the crystals that form.

Metal crystals

1 Set up the experiments as shown. (**Care**: The mercury can only be used in a demonstration by the teacher.)
2 Leave the crystals to form in a test tube.
3 Observe and draw.

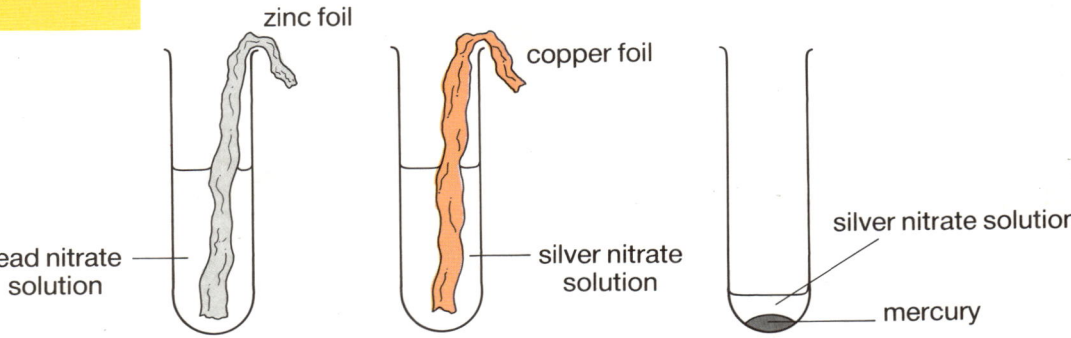

zinc foil

lead nitrate solution

copper foil

silver nitrate solution

silver nitrate solution

mercury

Examine the drawings that you have made.

Q1 Do different substances have the same shaped crystals?

Q2 Describe how you think the particles are arranged in each type of crystal.

Project

Find out about X-ray crystallography and what it is used for.

Structure and properties

Bonding compared

The following is a summary of the different types of bonding that can be found in everyday substances.

Ionic compounds ▶
Electrons are transferred between atoms forming cations and anions. These are strongly attracted to each other forming a giant crystal lattice. The ions can only vibrate, unless the solid is melted or dissolved in water when the ions become free to move in the liquid

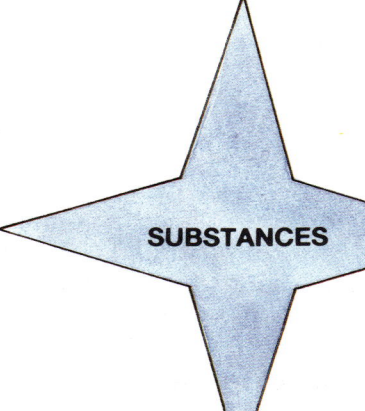

SUBSTANCES

Macromolecules ▲
These have giant lattices of atoms that are covalently bonded together

Metals ▲
These have positive metal ions closely packed together, surrounded by a sea of moving electrons

water

wax

Simple molecular compounds ▲
Atoms are covalently bonded together by sharing one or more pairs of electrons. The covalent bonds between atoms are strong but the forces between molecules are weak

Collect
- resource cards
- safety glasses
- summary table

1 Follow the instructions on each card, recording your results as you perform each task.

Take notes
Discuss your results with your partner and fill in the summary table.

Ions and electrolysis

It is possible to explain what happens during electrolysis in terms of ions and electrons.

Electrolysis of molten lead bromide

When lead bromide is melted and electrolysed the fumes evolved are very harmful. It is even dangerous to electrolyse lead bromide in a fume cupboard.

Study the diagram of the experiment in progress.

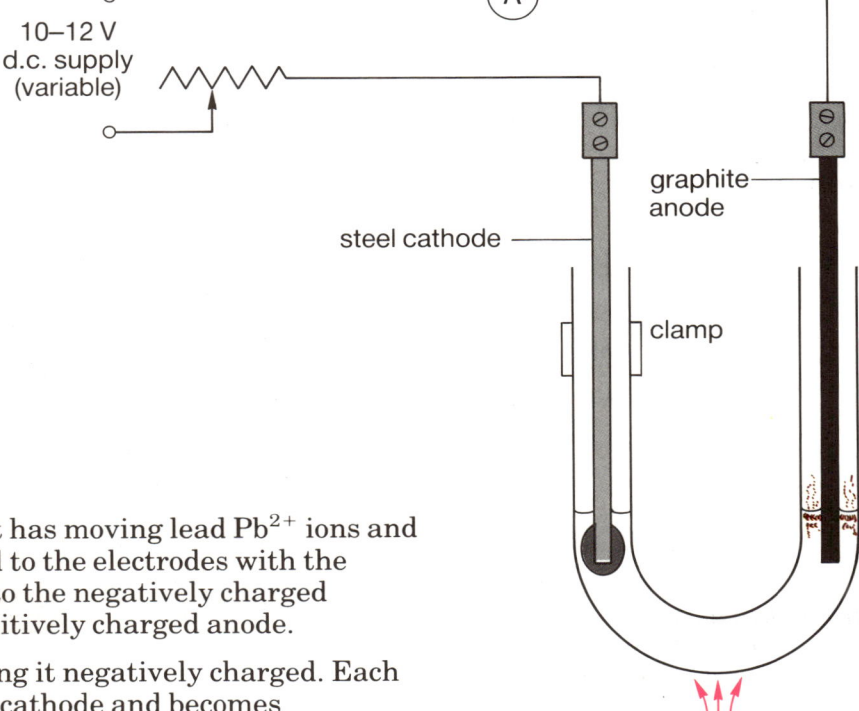

Q1 What can you see at the anode and at the cathode?

Q2 What do you think these products are?

When lead bromide becomes molten it has moving lead Pb^{2+} ions and bromide Br^- ions. These are attracted to the electrodes with the opposite charge. The Pb^{2+} ions move to the negatively charged cathode. The Br^- ions move to the positively charged anode.

The cathode has extra electrons making it negatively charged. Each Pb^{2+} ion gains two electrons from the cathode and becomes discharged as a lead atom.

$$Pb^{2+} + 2e^- \rightarrow Pb$$

Molten lead appears at the cathode.

The anode is short of electrons making it positive. Each Br^- ion gives one electron to the anode and is discharged as a bromine atom.

$$Br^- - e^- \rightarrow Br$$

Two bromine atoms join to form a bromine molecule.

$$Br + Br \rightarrow Br_2$$

Brown **bromine vapour** is seen at the anode.

Q3 Draw a labelled diagram of the experiment for electrolysing molten lead bromide. Show the movement of ions and what is formed at each electrode.

Q4 Predict what would happen during electrolysis of molten sodium chloride (NaCl).

> 📚 *Project*
>
> Find out what the industrial uses are for the electrolysis of molten sodium chloride.

Atomic models

One way of representing atoms and molecules is the three-dimensional model.

Design a way of drawing pictures of particles which enables you to show the structures listed.

(**Hint**:
  atoms could be a circle;
  different atoms could be different colours;
  ions are particles with positive or negative charges.)

Draw pictures to represent:

- a gaseous element
- a liquid compound
- a solid metal
- a mixture of gaseous atoms
- a solid which would conduct electricity when melted
- a pure compound whose molecules contain three atoms
- a solid which would have moving electrons in its structure
- a pure noble gas.

Talkabout

Removing electrons

Energy is required to remove an electron from an atom. It is called the **ionisation energy** and is measured in **kJ** (kilojoules). The table shows the energy needed to remove each electron in a certain atom.

Look at the shape of the graph. It represents the logarithm of the ionisation energy plotted against the number of ionisations. Using the logarithm of the ionisation energy makes it easier to find a suitable scale as, in this case, the successive ionisation energies can vary by several thousand kilojoules.

Electron	Ionisation energy (kJ)
1	500
2	4 600
3	6 900
4	9 500
5	13 400
6	16 600
7	20 100
8	25 500
9	28 900
10	141 000
11	158 700

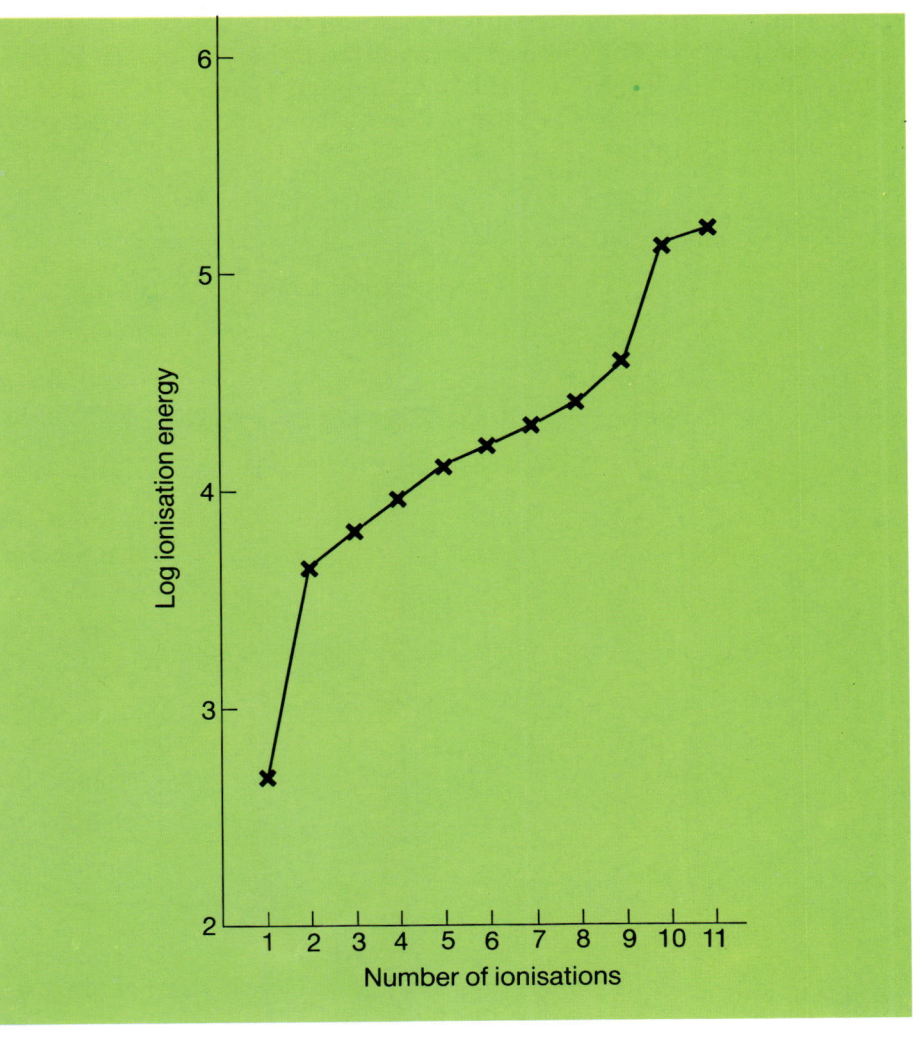

Work in a group. Collect a copy of the Periodic Table and discuss the following:

1 How many electrons are in the atom?
2 What element is composed of those atoms?
3 What connection can you see between the shape of the graph and the electron arrangement in the atom?
4 What shape might you get if a similar graph were drawn for the atoms of magnesium and boron?

Report your conclusions to the rest of the class.

1 The diagram shows some of the materials used for electric lighting. The wires which carry electricity to and from the bulb are insulated by the plastic, PVC. The filament is made from the metal tungsten. To prevent the hot tungsten from reacting with air the bulb is filled with an unreactive gas, argon.

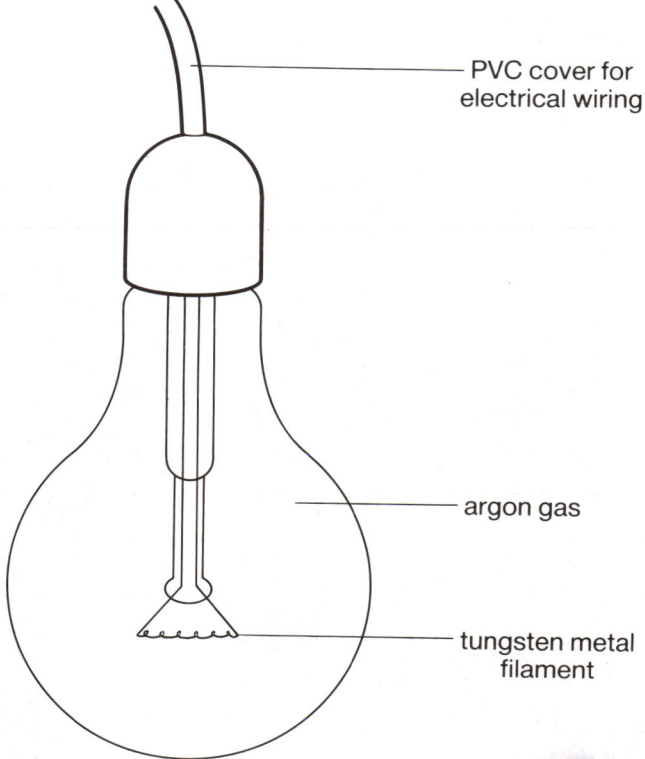

— PVC cover for electrical wiring

— argon gas

— tungsten metal filament

Explain why

a PVC is an electrical insulator [2]
b tungsten conducts electricity [1]
c argon is an unreactive gas. [1]

NEA 1989 Chemistry

2 a i Diamond and graphite are allotropes of carbon which both have giant molecular (macromolecular) structures. Explain in terms of their structures how the two allotropes differ. [2]
 ii State **one** physical property of diamond and graphite which is the same for each allotrope and **one** which is different. Explain in terms of their structures why the allotropes have the same property and why the other property is different for each allotrope. [4]

b i Describe and explain the different types of bonding which occur in the two compounds sodium chloride and hydrogen chloride. [5]
 ii State how you would expect the melting point of these compounds to differ, and explain your answer in terms of their bonding. [2]

NEA 1991 Chemistry

3 Draw a diagram to show the arrangement of all the outer energy level electrons in the atoms in a molecule of water. [2]

Name the type of bonding between the atoms in a water molecule. [1]

NEA 1988 Chemistry

4 The element gallium is in group III of the Periodic Table and exists as two isotopes $^{69}_{31}$Ga and $^{71}_{31}$Ga.

a Copy and complete the following table. [3]

	$^{69}_{31}$Ga	$^{71}_{31}$Ga
Number of protons		
Number of neutrons		
Number of electrons		

b State the number of electrons in the outer energy level of a gallium atom. [1]

c Write the formula for an ion of gallium. [1]

d Write the formula for
 i gallium chloride
 ii gallium sulphate. [2]

e A sample of gallium contains 60% of atoms of $^{69}_{31}$Ga and 40% of atoms of $^{71}_{31}$Ga. Which one of the following is the relative atomic mass of this sample of gallium? [1]

69.2   69.8   70.2   70.8

NEA 1984 Chemistry

Extensions

Litter pests?

Falling leaves settle on the soil. They do not stay there very long because they are broken down by invertebrates and further decomposed by fungi. These organisms are called **decomposers** and they perform the essential task of recycling chemicals from dead materials. The earthworm in the food chain below is a decomposer.

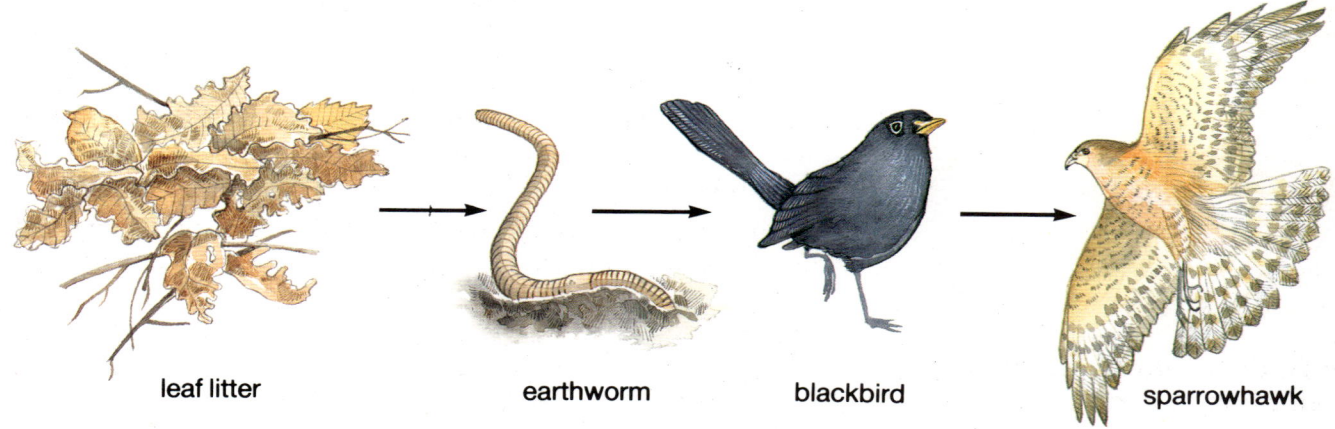

leaf litter   earthworm   blackbird   sparrowhawk

Collect

- leaf litter
- tray
- Tullgren funnel apparatus

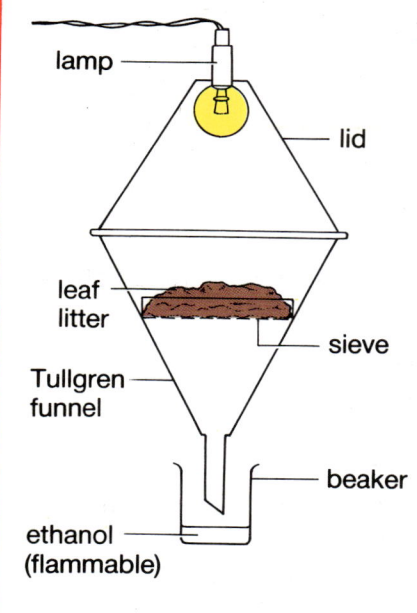

lamp

lid

leaf litter

sieve

Tullgren funnel

beaker

ethanol (flammable)

Litter logging

1 Collect some leaf litter from beneath a deciduous tree and weigh it.
2 Spread it on a tray and sort through it. Put larger animals and signs of fungi into a collecting dish.
3 Set up a Tullgren funnel as shown in the diagram. Leave for at least two hours.
4 Sort your collection into different feeding groups:

- Primary consumers eat the leaf litter and include: mites, millipedes, springtails, slugs, snails, earthworms and woodlice.
- Secondary consumers eat the primary consumers and include: beetles and their larvae, spiders, harvestmen, and centipedes.

Collect a resource sheet if you need help in identifying these organisms.

5 Count and record the number in each group, then weigh them.
6 Use these masses to construct a three-tier pyramid of biomass on graph paper. (How will you work out the mass of the producers?)

Q1 Explain why decomposers are essential in a food web.

Q2 Which organisms are likely to be present as decomposers but not included in your study?

Q3 What differences would you expect to find in a coniferous forest? Why?

Shrimps of the desert

Abiotic factors are very important to organisms. If these change, an animal or plant either adapts, moves on, or dies.

Brine shrimps once lived in pools in the salt works of Cheshire because this environment provided ideal conditions for them. However they are rarely found there nowadays because salt is no longer mined in Cheshire by pumping in water.

Brine shrimps can also be found in desert conditions. Deserts have high average temperatures (more than 18°C) and very little rain. During the day, temperatures can rise to above 40°C but the nights are much cooler. The little rain that falls runs over the surface and evaporates quickly.

The eggs of brine shrimps survive in the dry conditions. They hatch and reproduce quickly in the pools that form when it does rain.

How does light affect the behaviour of brine shrimps?

Using the illustration to help you, design an investigation into the reaction of brine shrimps to light.

Check your design with your teacher then carry out the investigation.

The behaviour patterns you have observed are **reflex actions**. These are simple unlearned responses which are inherited by the organism. They help an organism to survive by responding very quickly to environmental changes.

Q1 Describe how you carried out the investigation, using labelled diagrams.

Q2 Which variables did you control?

Q3 How could you improve your investigation?

Q4 Describe the effect of light on the behaviour of brine shrimps.

Q5 Describe any changes that took place on the surface of the shrimps (observed with a microscope).

Q6 What is a reflex action?

Q7 How do you think the behaviour of the shrimps assists their survival?

Copper—friend or foe?

Copper is a very important metal but sometimes the products of copper production escape into the water supply and seep into the nearby soil.

Bingham Canyon copper mine near Salt Lake City, USA

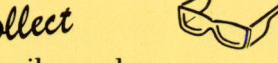

Collect

- soil samples
- flame-test equipment
- ammonia solution
- test tubes and rack
- copper sulphate solution
- quick-growing seeds
- petri dish
- safety glasses

How does copper affect plant life?

1 Find out if there is any copper present in the soil, by doing a flame test.
2 How much copper is present? When ammonia solution is added to a copper solution a clear deep blue colour is seen. Can you use this idea to find out how much copper is present? You may be able to use a colorimeter to measure how blue the solution is.
3 How does copper affect plant growth? You could use quick-growing seeds, such as cress, for this investigation. Remember to use a control in your investigation.

Q1 Imagine you are an environmental health official. Prepare a report about your findings on the effect of copper on plant growth.

[IT] Use a desk-top publishing program to prepare and edit your report.

Q2 You discover that the copper present in some soil samples has been released from a local factory. The factory wishes to increase its production. What will you recommend to the planning authorities?

Shadows

Shadows are formed because light travels in straight lines. The size of the shadow depends on how far the object and its shadow are from the source of light, and on the angle at which the light hits the object.

Investigate the size of shadows

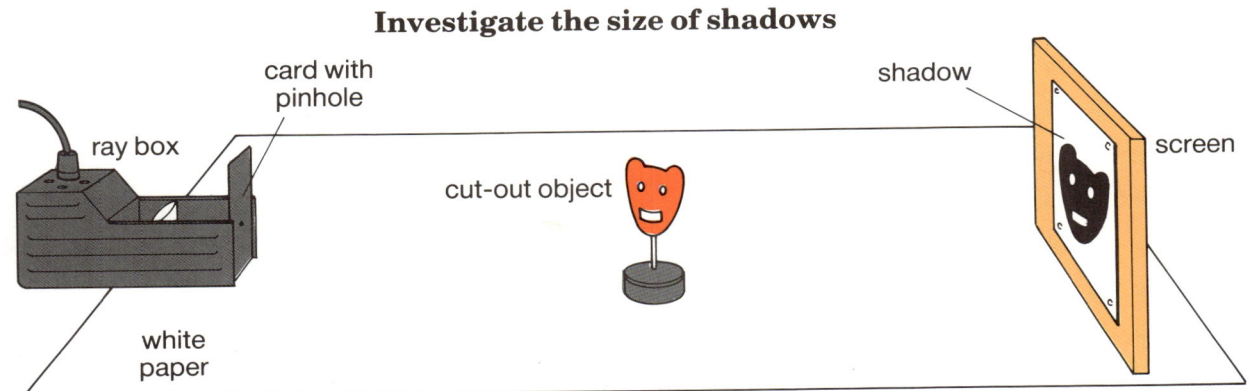

1 What do you expect will happen to the size of the shadow when you move the object towards or away from the ray box? Write down your ideas in a way which can be tested.
2 What measurements will you make? What will you keep constant?
3 How will you check that your measurements are accurate?
4 Carry out your investigation to see if your ideas are correct.
5 Look for a pattern in your results. Can you predict the height of the shadow without measuring it directly?

Q1 Write a report of your investigation.

Q2 Give your results in the form of a table.

Q3 What is the mathematical relationship between the size of the shadow and the distance of the object from the ray box?

Q4 The man forms a shadow 6 m high on the wall when he is 2.5 m from the floodlight. If he passed 4 m in front of the light how tall would his shadow be?

Optical instruments

Rays from an object close to a convex lens are refracted as shown left. The brain decides that the rays have travelled in straight lines from I, so we see the image there. This is called a **virtual image**. A **real image** can be projected onto a screen but a virtual one cannot.

Follow these ray diagrams which show how the image is produced in a microscope and in a telescope. Are the images real or virtual?

A magnifying glass is a simple convex lens which enlarges an object placed within its focal length

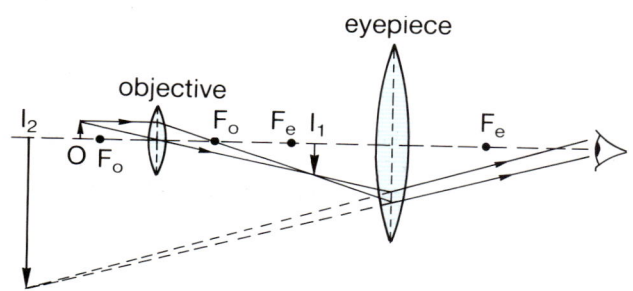

Microscope

Telescope

Make a telescope
Set up two lenses to investigate how a simple telescope works. As shown in the ray diagram above, the distance between the lenses must equal the sum of the focal lengths of the two lenses, $f_o + f_e$.

Look through the 'eyepiece' lens at an object across the classroom. You should see a sharp image. What do you notice about the image?

Swap the two lenses. Does the image change?

Try different combinations of lenses.

Make a table:

Focal length of eyepiece lens f_e	Focal length objective lens f_o	$\dfrac{f_o}{f_e}$	Magnification m

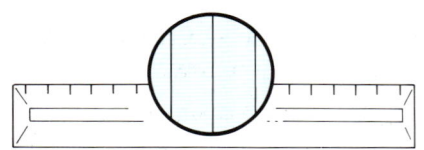

To work out the magnification of your telescope, position a ruler about 25 cm from the 'objective' lens. Look at the ruler through the eyepiece lens. Calculate the magnification:

$$\text{magnification} = \frac{\text{image size}}{\text{object size}}$$

Q1 Draw a labelled diagram of your telescope.

Q2 What was the best magnification you obtained?

Q3 What is the mathematical connection between the magnification and the focal lengths?

Interference and polarisation

Interference

When two or more waves cross the same path, the effects of the waves are merged or combined. You can see the effects of this superposition of waves with water. Next time you have a bath try creating waves. Watch what happens if you make two waves close together. You should see that they **interfere** with each other. The diagram below shows what happens.

The ripples interfere with one another and leave no ripple in some places

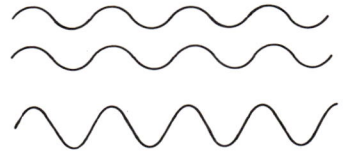

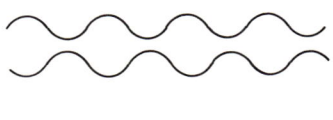

When waves of the same wavelength combine together they may produce a larger wave. This is called **constructive interference** and happens when the waves are 'in phase'

When the waves are 'out of phase' with one other they cancel each other out. This is called **destructive interference**

Polarisation

You may be familiar with the effect of polarisation by sunglasses.

'Polaroid' filters or coatings allow only about 30% of light rays to pass through and this is why they are used for sunglasses. The light which gets through is said to be **polarised**. If two Polaroid filters are placed on top of one another and one is rotated, there is a position in which no light can pass through at all.

Surfers are familiar with the effect of constructive interference when they wait for large waves to build up

Collect

- 2 cards and stands
- scissors
- thin string

Work in threes. Make slots in the two cards so that the string can pass through easily, and set up as in the diagram below.

Suggest a hypothesis about what will happen to the waves on the string as they pass through the two cards if the initial waves are in all directions.

What will happen if the second card is turned so the slot is at right angles to the other slot? What will happen to the waves as they pass through each slot?

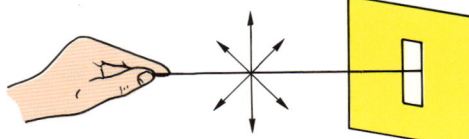
Vibrate the string in all direction

Q1 Make a report about your investigations.

Q2 How could your observations help to explain how light can be cancelled out with two 'crossed' Polaroid filters?

Q3 Find out what other uses are made of polarisation.

Further investigations of malachite

Your task is to find out as much as possible about the reactions of malachite. Remember the tests for negative ions. Use the data below to help you. The diagrams may give you some ideas.

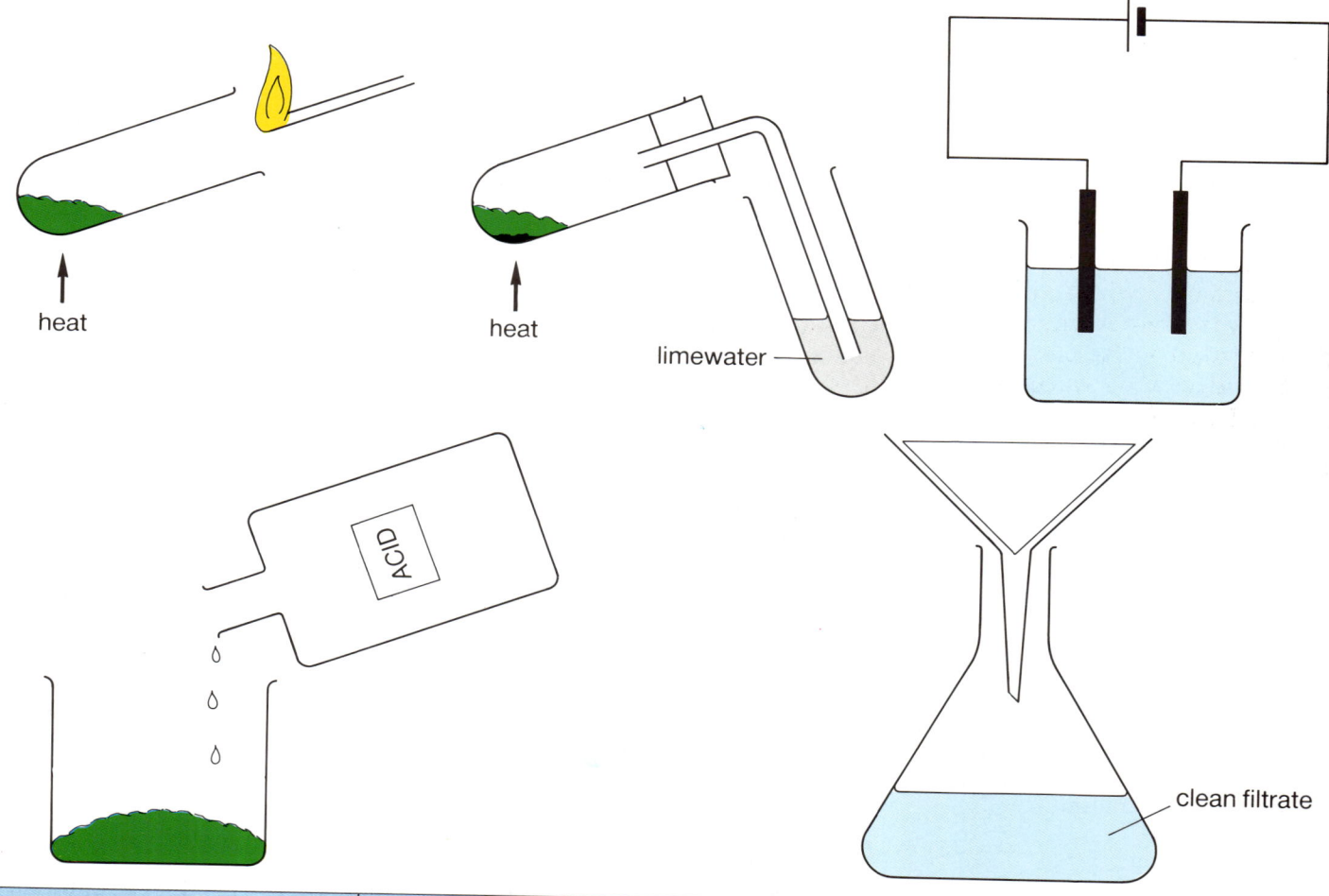

heat

heat

limewater

ACID

clean filtrate

Substance	Colour	Solubility	Electrical conductivity
Copper oxide	Black	Insoluble	Poor
Copper sulphate	Blue	Soluble	Good when in solution
Copper chloride	Blue/green	Soluble	Good when in solution
Copper nitrate	Blue	Soluble	Good when in solution
Copper	Pink	Insoluble	Good when solid or molten

Q1 Write a report detailing your experiments, observations and conclusions.

Q2 Try writing word equations for the reactions.

Electrolysis of dilute acids

This experiment will help you to understand more about the nature of dilute acids.

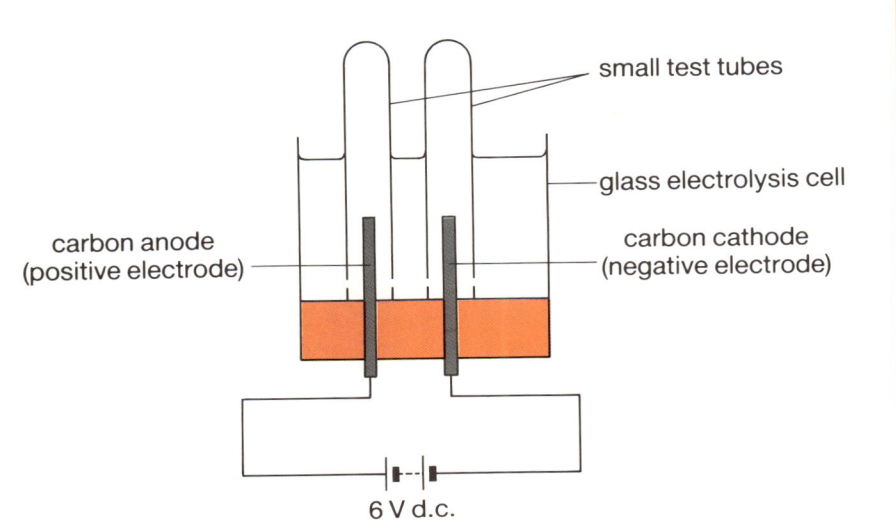

Collect

- electrolysis cell
- 2 test tubes
- 6 V d.c. supply
- dilute hydrochloric acid
- dilute nitric acid
- dilute sulphuric acid
- safety glasses

1 Set up the electrolysis cell and fill it with dilute hydrochloric acid.
2 Switch on the power supply.
3 Collect and test the gases given off.
4 Repeat using the other acids.

carbon anode (positive electrode)

small test tubes

glass electrolysis cell

carbon cathode (negative electrode)

6 V d.c.

You can use the following data to help you test the gases.

Gas	Colour	Test	Result
Chlorine	Yellow/green	Damp indicator paper	Bleaches
Oxygen	Colourless	Glowing splint	Relights
Hydrogen	Colourless	Lighted splint	Goes out with a squeaky pop

Q1 Draw up a table and complete:

Acid	Product at anode	Product at cathode

Q2 What pattern can you notice in your results?

Q3 If a solution conducts electricity what kind of particles—molecules, atoms or ions—must it contain?

Q4 What particle do you think is present in all the acids?

Q5 What particle do you think must be necessary for the acidic properties of the solutions?

179

Acid·indigestion

Some kinds of indigestion are caused by too much hydrochloric acid being produced in the stomach. There is a range of products on sale which can relieve the discomfort of indigestion. They all contain alkalis and they work by neutralising the acid.

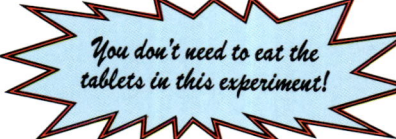

You don't need to eat the tablets in this experiment!

Design an experiment to find out which brand of indigestion tablet gives the best results

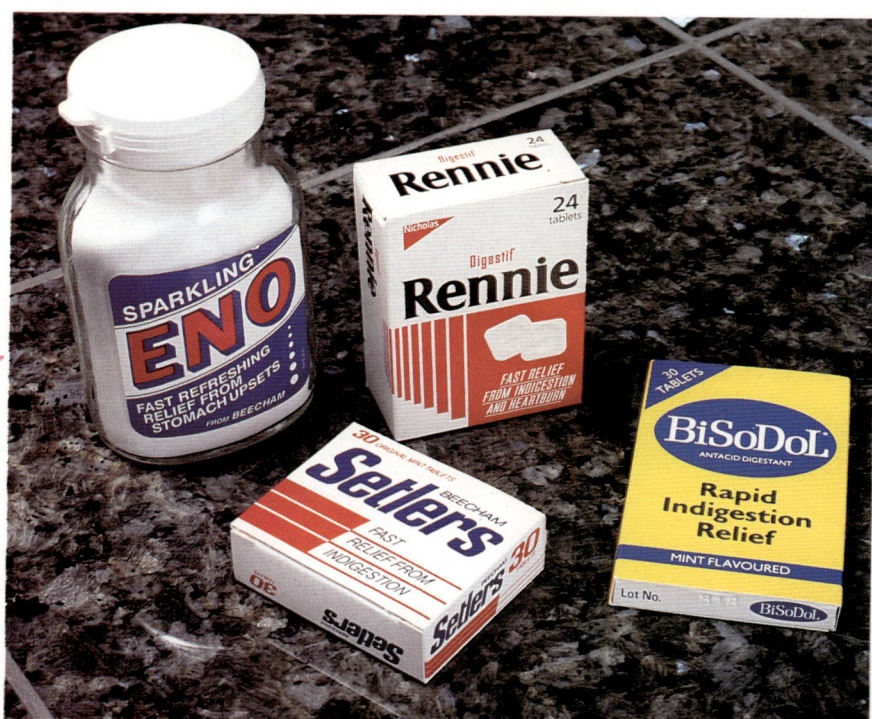

Things to consider:

- Do all the tablets react in the same way with acid?
- How will you know when neutralisation is complete?
- How will you make your tests valid? (There will be several variables to consider.)

Write an article for a health magazine, saying which indigestion tablet you would recommend as being good value for money. You should include how you did your tests and indicate how you made the tests fair.

Preparing a fertiliser

Ammonium sulphate is used as a fertiliser. Industrially it is made from sulphuric acid and ammonia. This is an example of **neutralisation** on a large scale.

$$\text{ammonia} + \text{sulphuric acid} \rightarrow \text{ammonium sulphate}$$
$$2NH_3(aq) + H_2SO_4(aq) \rightarrow (NH_4)_2SO_4(aq)$$

You can make ammonium sulphate yourselves.

Collect

- sulphuric acid
- ammonia solution
- universal indicator paper
- beaker (100 cm^3)
- beaker (250 cm^3)
- glass rod
- teat pipette
- watch glass
- evaporating dish
- bunsen burner, heat-proof mat, wire gauze, tripod
- measuring cylinder
- safety glasses

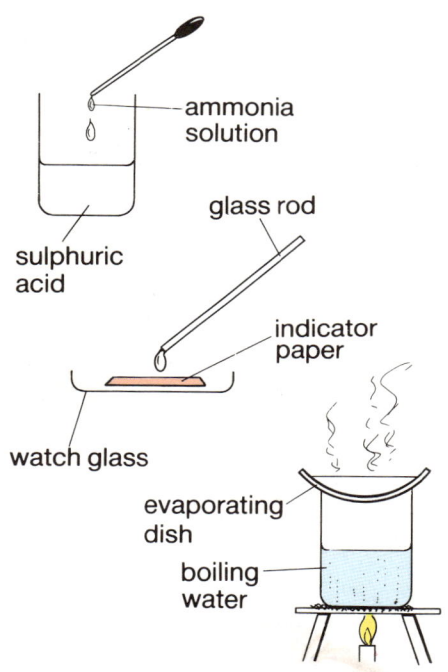

1 Pour 25 cm^3 sulphuric acid into the 100 cm^3 beaker.
2 Add a few drops of ammonia solution using a teat pipette and stir using the glass rod.
3 Take out a drop of the solution using the glass rod and dab it onto universal indicator paper.
4 Repeat the above three steps until the indicator paper turns blue. What does this colour indicate?
5 Pour the solution into an evaporating dish. Put the dish carefully on a large beaker of boiling water and heat until the volume has decreased by half.
6 Remove from the water bath and leave to cool.

Q1 Discuss with your partner how you would find out if ammonium sulphate improves the growth of plants from seed. Write down your plan. Include:

- which seed you would use
- what containers you would use
- what growing medium you would use, e.g. soil, cotton wool
- how you would make your tests valid
- how you would measure growth.

Q2 Include in your plan an investigation of how, if at all, different concentrations of ammonium sulphate affect plant growth. Write down how you would do this. Remember to include methods to make your results valid.

Check with your teacher and then carry out your investigation planned in question 1. If you have time, carry out the second investigation too.

Investigating the effect of light on a chemical reaction

Have you ever noticed that colours in the fabric of window blinds and deck chairs tend to fade after a long time? This is because the sunlight causes a bleaching reaction. There are many chemical reactions that take place only in the presence of light. Can you think of other everyday examples?

Collect

- filter paper
- filter funnel
- bottle of distilled water
- measuring cylinder
- beaker
- glass rod
- silver nitrate solution
- potassium bromide solution
- conical flask
- coin (1 pence piece)
- spatula
- safety glasses

Investigation

1 Mix 10 cm^3 silver nitrate solution and 10 cm^3 potassium bromide solution in a beaker and stir.
2 Filter off the precipitate and wash it with distilled water.
3 Spread the precipitate on the filter paper and place a small coin over it.
4 Put the paper on a window ledge and leave for 15 minutes.
5 Remove the coin. What has happened?

The silver bromide which was made during the reaction decomposes slowly to silver which causes the darkening effect. Light, like heat, is a form of energy and speeds up the reaction. Therefore the covered part of the precipitate remains paler.

Q1 Find out how an image is produced on photographic film.
Q2 Find out about another light-sensitive reaction, for example photosynthesis.

Detecting charge

When electrons are transferred from the surface of one object to the surface of another object, the objects become oppositely charged with **static electricity**.

A **gold leaf electroscope** is a simple instrument that uses **repulsion** to detect charge.

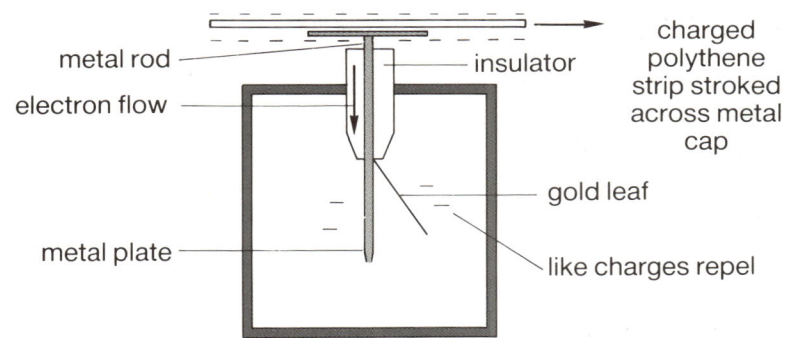

metal rod

electron flow

metal plate

insulator

charged polythene strip stroked across metal cap

gold leaf

like charges repel

Make your own electroscope

Collect
- metal can
- Blu-Tack
- pins
- thin silver paper, e.g. from chocolate
- plastic container, e.g. yoghurt pot
- rice

1 Set up your electroscope as in the illustration above.
2 Charge it by shaking rice into it.
3 Make some hypotheses and then test them:
   a What do you expect to happen when you charge another object and then put it inside the can?
   b What will happen when you put a charged object close to the 'arms' of your electroscope?
   c Can you tell whether an object has the same charge as your electroscope, or the opposite charge?
   d Can you measure the charge on the objects?
   e How can you remove the charge?

Write a report about your electroscope. Remember to include a method section with diagrams, results and whether or not your hypotheses were correct.

Can you suggest how you could improve the design of your electroscope?

Household wiring

However large or small the building you live in, it will have been wired following strict rules. It is important that these rules are followed carefully by qualified electricians to avoid the dangers of electric shocks and fires in the home. The main parts of any household wiring system are as follows.

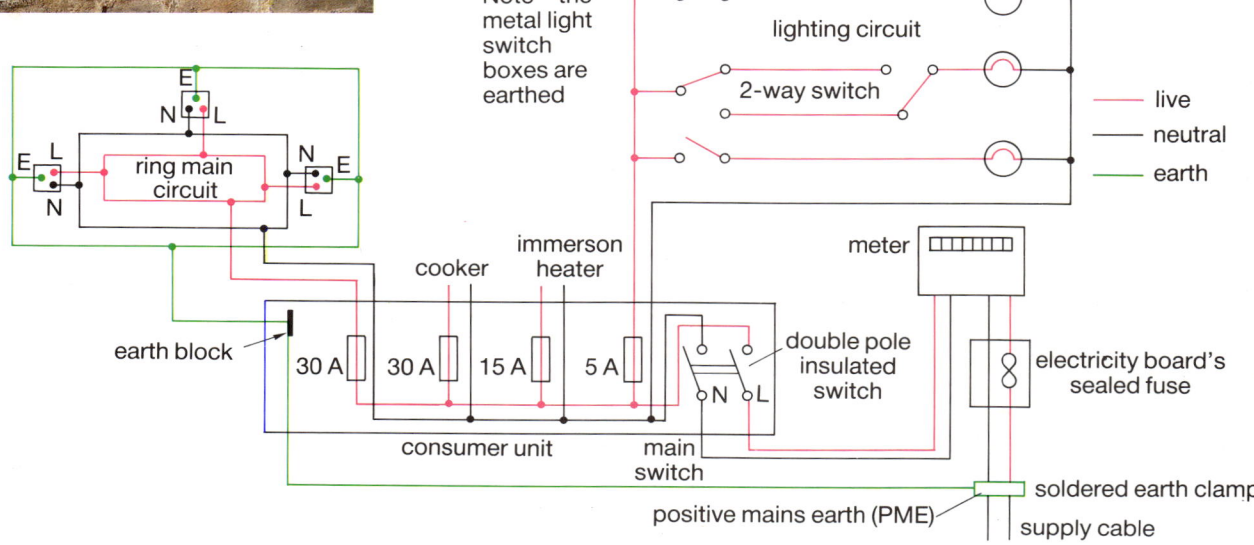

Where the mains electricity cable comes into your house, the live wire passes through the mains fuse which you are not allowed to touch. If there is a serious electrical overload on the house wiring, this fuse will 'blow' to protect the house from fire.

All the electricity passes through the meter where the current used is recorded, then the live wire splits into several wires, each of which passes through a fuse in the fuse box. These are linked to various circuits in your home. The main ones are usually:

- upstairs lights
- downstairs lights
- upstairs sockets
- downstairs sockets
- electric cooker
- immersion heater(s)

The supply voltage on each circuit remains fairly constant at 240 V.

Each circuit has a specific thickness of wire and size of fuse in the box, appropriate to the power used by the appliances in the circuit. It is very important that the correct size of fuse is used when making a replacement and that the correct thickness of cable is used when wiring up devices.

Q1 Why does an electric cooker have a circuit of its own?

Q2 If a 15 amp fuse has blown in your fuse box, you should of course replace it with another 15 amp fuse. What problems might there be if you replaced it with
  a a 5 amp fuse?
  b a 30 amp fuse?

Study the diagram of a house wiring system.

Q3 Why are the lights in parallel with each other?

Q4 How does the two-way switch work?

Q5 Try to find the meaning of the following terms: *junction box, circuit breaker, residual earth leakage*. These things can all be found in wiring systems.

The National Grid system

The National Grid in England and Wales is a system for distributing electricity. It is supplied with electricity generated by a variety of companies. These include *National Power*, *PowerGen* and *Nuclear Electric*. In Scotland the main companies, *Hydro Electric* and *Scottish Power*, are responsible for both generating electricity and distributing it.

The electricity from power stations is distributed via a network of cables on the National Grid, mostly supported by pylons

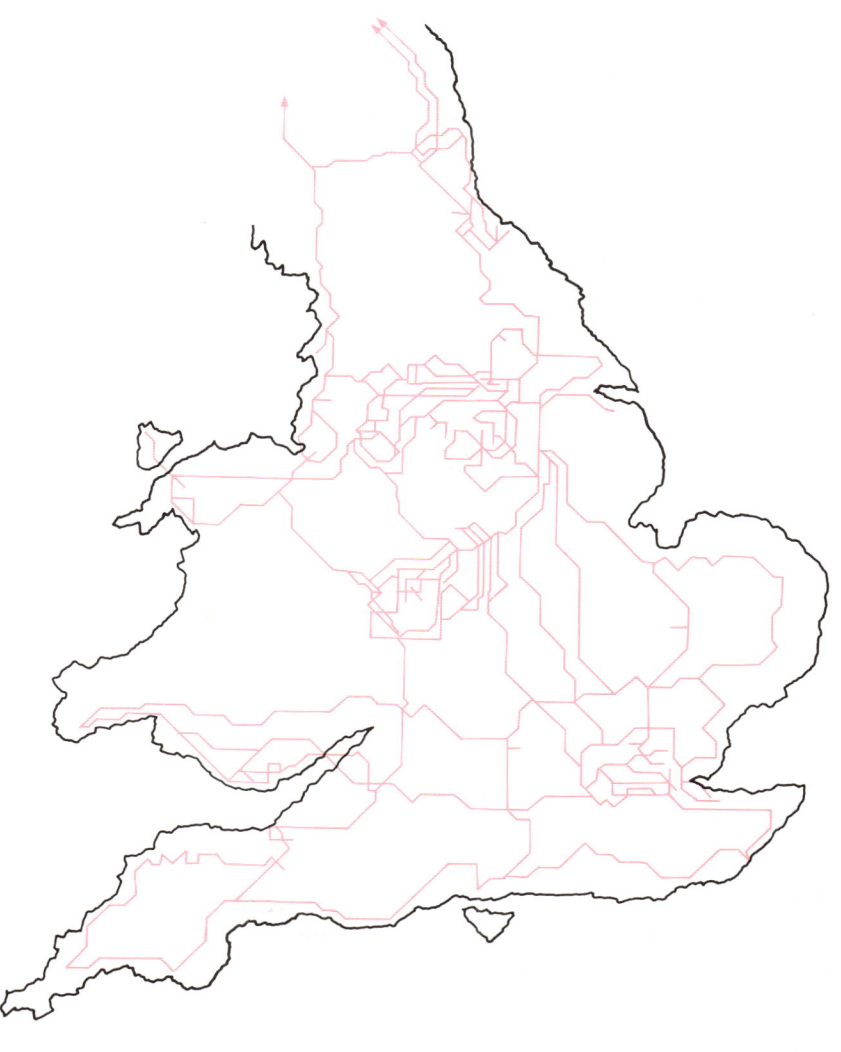

1 Collect jigsaw pieces.
2 Assemble the jigsaw and stick it in your book.
3 Use a coloured pencil to connect the parts labelled A–N (in alphabetical order).

Q1 What kind of building receives 33 000 V?

Q2 What voltage do homes receive?

Q3 What purpose do you think the transformers have?

Q4 What purpose do you think the circuit breakers have?

Q5 What do you think the dotted lines represent?

Relays

A relay is a switch controlled by an electromagnet. It allows a circuit with a small current to control one with a large current.

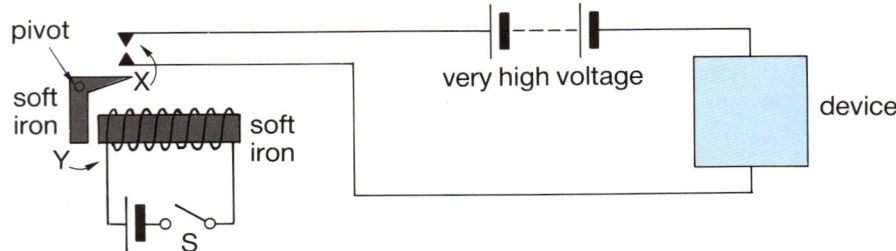

Closing the switch S makes an electromagnet which attracts the hinged shapes at Y. This pushes X up which closes the point contacts, completing the high-powered circuit.

Relays have many varied uses, e.g. in telephone circuits, in car ignition circuits.

A relay switch from a car ignition

Collect
- connecting leads
- switch
- two batteries
- wire
- soft iron core
- magnetic switch
- two bulbs

Set up this pair of circuits.

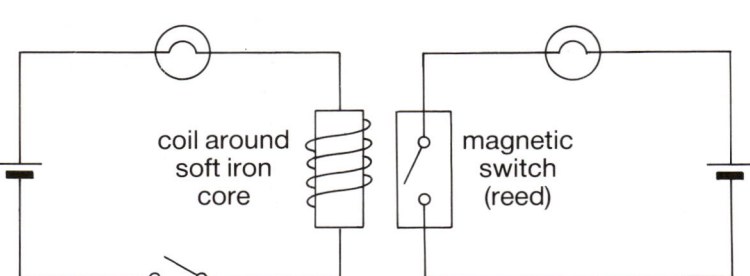

Close the switch in the first circuit and note what happens in the second circuit.

Q1 Write a report of your experiment.

Q2 Give two possible advantages of switching on a circuit in this way.

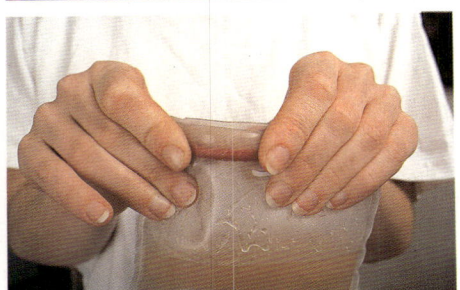

Handwarmers

You can buy chemical handwarmers from outdoor pursuit shops. There is one kind that is reusable. It uses the chemical called **sodium ethanoate**. The handwarmer is a sealed plastic bag containing **hydrated** sodium ethanoate. This means that when the sodium ethanoate solidified, molecules of water were 'built into' the crystal. This is called **water of crystallisation**.

To get the handwarmer ready for use, you heat the whole bag in a pan of boiling water until the sodium ethanoate is completely liquid. It does not melt, it just dissolves in its own water of crystallisation. After the handwarmer has cooled down, it can be packed in a rucksack. When it is needed, you bend a small metal plate inside the bag that 'clicks' sharply. This causes the sodium ethanoate to crystallise by 'jarring' the solution. The heat put in (to form the solution) was stored and is now released when the solution crystallises.

Try making your own test-tube version using the same chemical.

Collect

- large test tube and rack
- 250 cm³ beaker
- hydrated sodium ethanoate
- thermometer
- bunsen burner
- tripod and gauze
- heat-proof mat
- safety glasses
- stirring rod
- teat pipette
- stopclock
- distilled water

Handy handwarmer

1 Half fill a test tube with sodium ethanoate crystals.
2 Put the test tube in a beaker of boiling water.
3 Wait until all the sodium ethanoate has dissolved. If the last few crystals do not seem to be dissolving, add one or two drops of distilled water, making sure that you wash any chemical sticking to the sides of the test tube down to the bottom. This is called a **saturated solution**.
4 Take the test tube out of the water and let it cool down. It should still be a solution. This is called **super-saturated solution**.
5 Put in a thermometer. 'Flick' the test tube sharply to make the super-saturated solution crystallise. If this does not work, add a tiny crystal of sodium ethanoate to act as a 'seed'. Take the temperature every five seconds.
6 Check to see if the chemical *is* reusable by heating it up and crystallising it again.

Q1 Plot a graph of temperature against time.

[IT] Use a data-handling program to plot the graph.

Q2 Make a report of the experiment. Include with the report your graph and a labelled energy level diagram of the reaction.

Hydrogen as a clean fuel

Hydrogen gas is an element. It consists of molecules made up of two hydrogen atoms (H_2). Hydrogen can be produced when electricity is passed through water with acid or some kind of salt dissolved in it to improve the conductivity.

Hydrogen combines readily with oxygen when ignited and produces water and releases a great deal of energy.

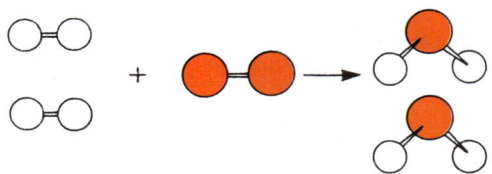

The energy released when hydrogen combines with oxygen can be harnessed in two ways:

1 Burning a mixture of hydrogen and air and trapping the heat.

2 Allowing hydrogen and oxygen to produce an electric current by combining slowly in a fuel cell.

Collect

- electrolysis apparatus
- small test tubes
- power supply (6 V d.c.)
- leads and crocodile clips
- voltmeter
- sodium hydroxide solution
- wooden splint
- safety glasses

1 Electrolyse some water to produce hydrogen and oxygen.

2 Test the gases by placing a burning splint and separately holding a glowing splint in the mouth of each of the test tubes. Work out at which electrode hydrogen is formed. Try to remember the tests for oxygen and hydrogen.

3 Set up the apparatus again and produce full test tubes of hydrogen and oxygen. You will probably have to collect two tubes of hydrogen to get one of oxygen.

4 Disconnect the power supply and connect a voltmeter. What reading do you get?

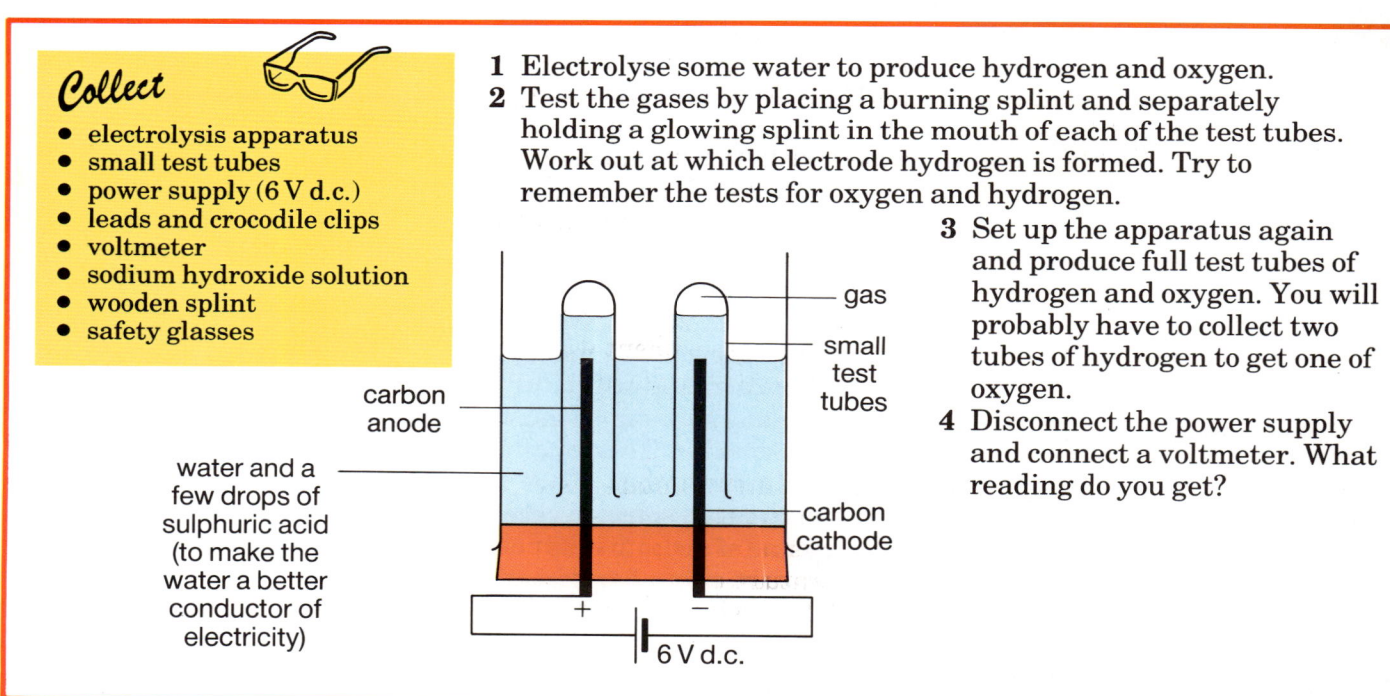

carbon anode

water and a few drops of sulphuric acid (to make the water a better conductor of electricity)

gas

small test tubes

carbon cathode

+ − 6 V d.c.

Q1 Write a report of your experiments. Include:

a at which electrode hydrogen is produced

b what you observe when hydrogen is burned

c the product of your reaction(s).

Q2 Is hydrogen a good fuel? Explain your answer. (The heat of combustion of hydrogen gas is about 140 000 kJ per kg.)

Larger molecules—more heat?

The first four members of the hydrocarbon series are methane, ethane, propane and butane. There is a similar set of compounds called alcohols. The first four members of this series are:

methanol CH_3OH ethanol C_2H_5OH propanol C_3H_7OH butanol C_4H_9OH

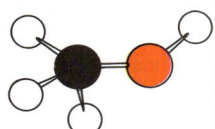 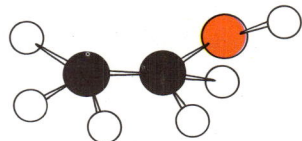 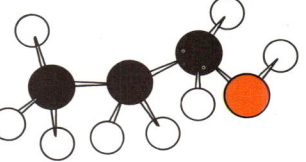

Each of the alcohols is a liquid. Each has one extra carbon atom and two extra hydrogen atoms than the one before it in the series. They all burn completely in oxygen to form carbon dioxide and water.

Each member of the alcohol series has an extra

$$\sim \overset{\displaystyle H}{\underset{\displaystyle H}{C}} \sim$$

than the member before it in the series

Collect

- thermometer
- metal can
- alcohol burner
- methanol/ethanol/propanol/butanol
- tripod and gauze
- heat-proof mats
- measuring cylinder
- safety glasses

Does adding an extra $-CH_2-$ increase the amount of heat that can be released?

1 Set up the apparatus. Start with methanol.
2 Note the weight of the burner and temperature of the water.
3 Light the burner and heat the water until its temperature rises by about 30 °C. Note the exact temperature. Put the cap back on the burner.
4 Weigh the burner (with the cap on).
5 Repeat the experiment with at least two more alcohols.

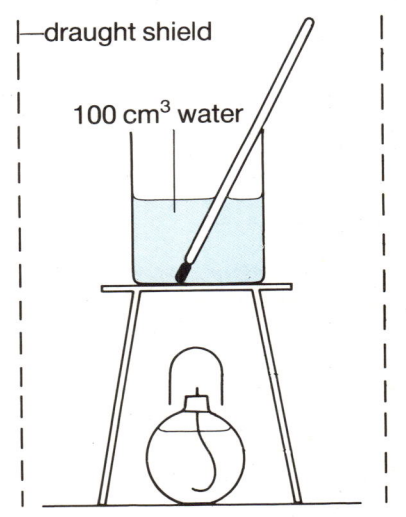

draught shield

100 cm³ water

Q1 Write a report of your experimental method and tabulate your results.

In order to compare the amount of heat supplied by the different alcohols, we need to find out what the temperature rise would be if we burned the *same number* of molecules of each alcohol.

Q2 Use a calculator to do the following calculation for each alcohol.

$$\frac{\text{mass of alcohol (g)}}{x} \times \frac{\text{temperature rise}}{\text{(°C)}} =$$

Write down this new figure for the temperature rise for a standard number of molecules of each alcohol in a table.

Q3 Draw a bar chart of number of carbon atoms in each alcohol molecule against the temperature rise (for standard number of molecules). Interpret your chart.

	x
methanol	0.32
ethanol	0.46
propanol	0.60
butanol	0.74

Messages from space

One important modern energy changer is the radio telescope. Optical telescopes simply magnify the light images received from space but radio telescopes detect radio waves which can be changed into images on a screen. They can observe events happening in far distant parts of the Universe.

In 1931 radio waves coming from space were discovered by accident. An American radio engineer, Karl Jansky, was studying interference in shortwave radio communications. He discovered that a source of the interference was radio waves coming from the constellation of Sagittarius—far beyond the solar system.

After the Second World War radio telescopes were built in many parts of the world and astronomers discovered that large numbers of objects, from galaxies down to planets, emit radio waves.

Using radio telescopes it is possible to observe parts of the Universe so distant that the radio waves have taken thousands of millions of years to reach us and we may be observing events that happened close to the beginning of the Universe.

In the 1960s radio astronomers discovered new objects called **quasars** and **pulsars**. Quasars (or quasi-stellar radio sources) are very distant astronomical objects which appear in photographs as extremely bright stars (as bright as all the light from a whole galaxy together). The nature of quasars is still being debated but it is thought that the intense radiation emitted may be due to stellar material falling into a 'black hole' at the heart of a distant galaxy. The lost gravitational energy becomes radiated electromagnetic energy; for example, visible light and radio waves.

In 1967 sources of radio waves emitting rapid pulses of radio energy were detected by the Radio Astronomy Group at Cambridge and named pulsars. Pulsars are thought to be rotating **neutron stars**—small, extremely dense remnants of supernova stellar explosions within our galaxy.

Collect a copy of the newspaper article and explain why it was described as 'sensational news'. Why is it unlikely that the new planet could sustain life?

Very Large Array of radio telescopes in New Mexico, USA

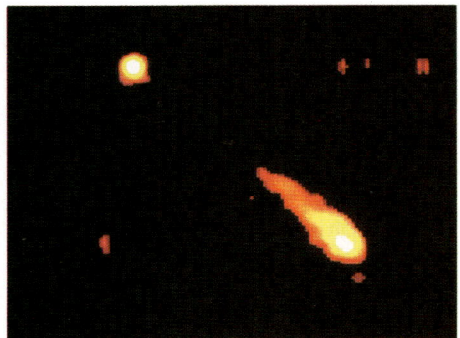

Radio image of quasar 3C 273 and its jet (bottom right), also a powerful radio emitter

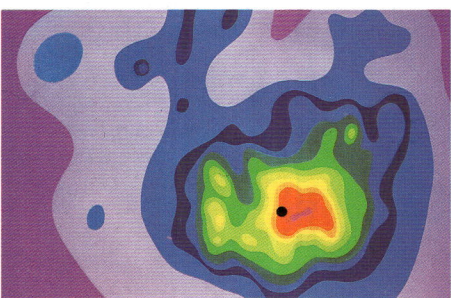

False-colour radio image of the Vela supernova remnant. At the centre (shown by a black dot) is the Vela pulsar which spins 13 times a second emitting pulses of light and radio waves

Q1 The space age has brought other ways in which we can investigate both the solar system and beyond. Find out what these are and what the advantages are. (Think about atmospheric and pollution effects which affect observation of the night sky.) A good astronomy book will provide plenty of ideas.

Q2 Find out about the constellation of Sagittarius. What does the name mean? Where have you heard it before?

Q3 What energy change is thought to take place within a quasar?

Save energy

We are constantly reminded to save energy. Often we think it is not worthwhile going round the house switching off lights or reducing the temperature by one degree as we are advised. However, it is like voting; if everyone does it it has an enormous impact.

This article is from the *New Scientist* of 5 August 1989. Read it carefully, then answer the questions.

Energy inefficiency brings a warm glow to the welcome in Budapest

STEPHEN Lindsay, an energy management consultant, received a particularly warm welcome in Hungary three winters ago: his first appointment at a hospital complex in Budapest took him to the director's office where the temperature was 32 °C, nearly twice what it needed to be and not far short of body temperature. Snow, about 600 millimetres deep, lay on the ground outside and all the windows of the room were wide open, recalls Lindsay. "The director was concerned about energy consumption. He told me: 'Our X-ray equipment is simply devouring electricity. Can you help?' Despite the fresh snow, I surveyed the entire complex without getting my shoes wet by simply following the routes of the underground heating mains, none of which was insulated." It turned out that 90 per cent of the hospital's annual energy bill of £500 000 went on heating. Things have changed little since then.

The waste of energy at the hospital is typical of what goes on elsewhere in Hungary, particularly in industry, and in Comecon states generally. The root of the problem is a lack of accountability. For example, factories in Hungary often take steam from a neighbouring factory when they could raise steam more efficiently themselves. Hungarian industry is centralised with energy consumption dominated by about 165 companies. According to Lindsay, even a rudimentary programme of energy conservation that concentrated on these companies could reduce consumption by 25 per cent. This would mean coordinating their activities; at the moment, no mechanisms exist to forge a joint approach to conservation. Also there are few people in the country with the necessary skills. To comply with regulations, companies nominate one of their employees to monitor energy consumption at large industrial sites. These impromptu "energy managers" rarely have enough power, expertise or resources to do the job properly.

The result is that opportunities to make large savings of energy are wasted. Pipes

Poor record: cheap fuel and extravagant power generation in East Germany have created polluted cities and dying forest

that could recycle steam from a generator to a boiler have rusted through; patching the holes at a cost of £50 would save £30 000 per year. Broken valves cause steam and hot water to run through mains, which are usually unlagged, 24 hours a day, seven days a week. A pressurised circuit in a chemicals factory or in a power station that is designed with three compressors will very soon need a fourth to compensate for leaks.

Hungary's greatest folly, however, was to become involved in the Danube hydroelectric scheme ("Hungarian Greens with the Danube blues", *New Scientist*, 18 August 1988). This involves the construction of two power stations, one at Gabcikovo in Czechoslovakia and another, now almost certain to be cancelled, at Nagymaros in Hungary. The cost of the project, which is being shared with Czechoslovakia, is more than £1 billion. But what about the benefits? Hungary is due to receive half of the power generated; this will be about 1800 million kilowatt-hours, just 4 per cent of the country's demand. But first it must reimburse Austria, which financed the scheme: for the first 20 years that the scheme operates, two-thirds of the energy due to Hungary must be exported to Austria. In winter, Austria must receive twice as much power as it does in summer, which requires Hungary to build more coal-fired power stations to generate electricity for export. No wonder Hungary wants to withdraw from the scheme. A programme of energy conservation in the first place would have saved a lot of trouble.

Ivan Vince

Q1 Make a list of simple and inexpensive things which could help save energy in Hungary.

Q2 How are energy managers chosen and why can't they do their jobs properly?

Q3 Why was the involvement in the Danube hydroelectric scheme described as Hungary's greatest folly?

Q4 What would your advice be to the Hungarian government to help with improving energy efficiency?

Q5 Design a poster for the government to circulate round factories and hospitals to encourage energy saving.

Ebb and flow

Tidal effects are mainly due to the moon's gravitational pull on the Earth. This results in a tidal bulge on the side of the Earth facing the moon and a similar one on the opposite side of the Earth.

As the Earth rotates, two high tides and two low tides pass each tidal point in 24 hours and 50 minutes. This means that high tide is at a different time at different spots around the coast and a slightly later time each consecutive day. It is important to know this if you are a fisherman or are on holiday and wish to swim.

Twice a month there are very large tides. These are called **spring tides** (which are named after a leaping movement not the season). The twice-monthly small tides are called **neap tides**.

These variations are due to different alignments of the Earth, the moon and the sun as the moon orbits the Earth. The gravitational pull of the Earth and sun contribute to the tidal effect. There are also seasonal variations.

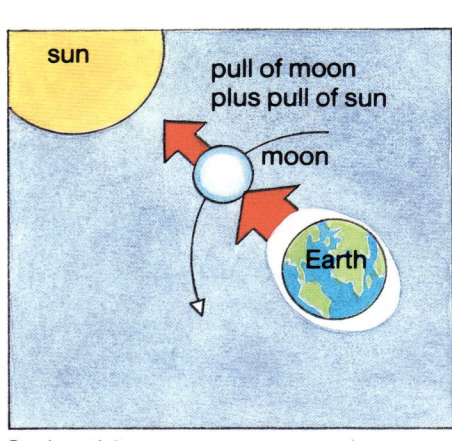

Spring tide

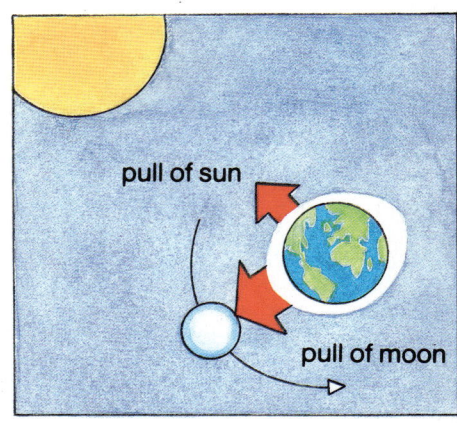

Neap tide

The following table shows the variation in the high tide level on the River Thames at London Bridge.

Date:	1	2	3	4	5	6	7	8	9	10	11	12	13	14	15	16	17	18	19	20	21	22	23	24	25	26	27	28	29	30	31
															January																
Ht of tide/m	7.5	7.5	7.3	7.1	6.8	6.6	6.4	6.3	6.3	6.5	6.7	6.0	6.9	7.0	7.0	7.0	6.9	6.8	6.6	6.4	6.2	6.0	5.8	5.8	6.2	6.6	6.9	7.1	7.4	7.5	7.6
Lunar phase					◗										◯						◗							⬤			
														February																	
Ht of tide/m	7.6	7.4	7.1	6.7	6.4	6.1	6.0	6.2	6.6	6.6	6.8	6.9	7.0	7.1	7.1	7.0	6.8	6.7	6.5	6.2	6.0	5.9	6.1	6.6	6.8	7.1	7.4	7.5			
Lunar phase				◗								◯								◗							⬤				
												March																			
Ht of tide/m	7.7	7.6	7.5	7.1	6.8	6.4	6.0	5.8	6.0	6.3	6.4	6.7	6.8	7.0	7.1	7.2	7.1	7.1	6.9	6.7	6.4	6.1	6.0	6.2	6.7	7.1	7.1	7.3	7.4	7.6	7.6
Lunar phase				◗									◯								◗								⬤		

Q1 Plot a graph of the variation in high tide level at London Bridge.

Q2 On your graph mark the positions of the spring and neap tides.

Q3 What phase is the moon in for each of these tides?

Q4 Why is it important to know when the tides are going to be high? What precaution has been made on the Thames?

Q5 Tidal energy can be used to generate electricity. Find out about the proposed tidal power station in the Severn Estuary and prepare a report to be presented to the class.

Falling off a cliff

It takes a force to start things moving, to speed them up, to slow them down, to change their direction, and to stop them. Everyone has seen cartoons like the one shown where the cartoon character carries on going in a straight line instead of falling down as soon as the edge of the cliff is reached. Does the direction in which the object is moving as it goes over the edge affect the time taken to reach the bottom of the cliff? What happens when the object is moving horizontally and what happens when it is moving vertically?

Collect
- ruler
- 2 coins

1 Take two identical coins and position them as shown in the diagram.

Give the ruler a sharp tap so that both coins jump off together.

Listen carefully. Do the coins land together?

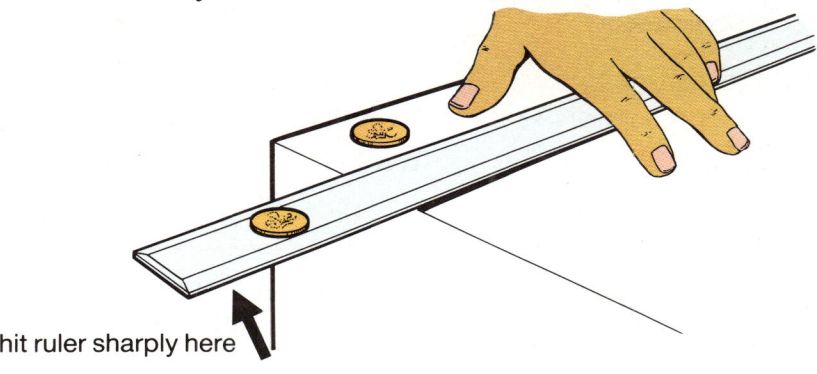

hit ruler sharply here

Collect
- 2 tennis balls

2 Work in a group so that some of you can observe carefully. Make sure that you have enough space to carry out the activity.

One person should take a ball in each hand. Throw one sideways at the same time as dropping the other one. Do they hit the ground together?

Q1 Sketch and label the paths of the two coins in experiment 1.

Q2 Describe what happened in the second experiment.

Q3 Look at the picture of the skaters. Which will have the greater acceleration. Why?

How efficient is your kettle?

When you boil water in a kettle, not all the heat goes into the water. Some heat energy is lost to the surrounding area. This means that energy is wasted and the efficiency of the appliance is reduced.

$$\text{Efficiency} = \frac{\text{useful energy transferred (output)}}{\text{total energy transferred (input)}} \times 100\%$$

The **specific heat capacity** of water is the heat energy that has to be transferred to 1 kg of water to raise its temperature by 1 °C. This is 4200 joules and we can use this to work out the useful energy transferred.

$$\underset{\text{(J)}}{\text{Useful energy transferred}} = \underset{\text{(kg)}}{\text{mass}} \times \underset{\text{(J/kg/°C)}}{4200} \times \underset{\text{(°C)}}{\text{temperature rise}}$$

The mass of water is simple to calculate if the volume is measured, because the mass of 1 cm³ is 1 g.

The power of the kettle will be marked on a plate on the side or the base. This is often 2.5 or 3 kW (kilowatts). One kilowatt supplies one kilojoule (1000 J) of energy every second. So

$$\underset{\text{(J)}}{\text{Total energy transferred}} = 1000 \times \underset{\text{(kW)}}{\text{power of kettle}} \times \underset{\text{(s)}}{\text{time}}$$

Why should you not boil half a kettle for one cup of tea?

Collect

- kettles
- measuring jug
- thermometer
- stopclock

Compare the efficiency of a traditional metal kettle and a plastic jug kettle

Plan and carry out the investigation. You will need to:

- Work out the energy transferred to boil some water in the kettle. Remember water boils at 100 °C.
- Work out the electrical energy transferred from the mains.
- Calculate the efficiency of your kettle.
- Repeat with different amounts of water. Do you get the same value for the efficiency each time? If not can you explain the differences?
- Repeat for the other kettle and compare your results.

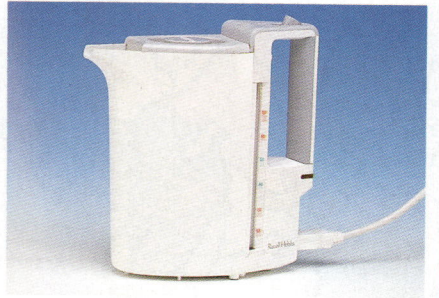

Q1 Write a report on your investigation. If efficiency were your main concern, which kettle would you choose if you wanted to buy a new one?

Are you an owl or a lark?

Some people's temperature varies regularly over a period of twenty-four hours. People who have their lowest temperature late in the evening tend to prefer getting up early and feel tired later in the day ('larks'). On the other hand those whose temperature peaks late in the day often feel sluggish and dazed in the morning but feel active and willing to do things late at night ('owls'). These rhythms are very difficult to alter even if you change your sleeping/waking habits. This is an example of a **circadian rhythm** (circadian means around one day).

Q1 Plot these figures on graph paper. They show how Chris's mouth temperature varied over a period of two days.

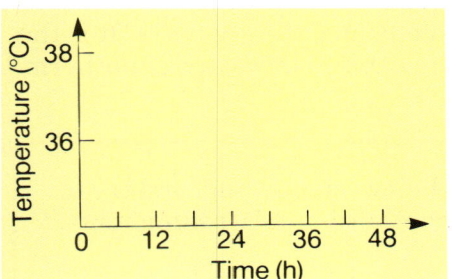

[IT] Use a data-handling program to plot the graph.

THURSDAY	24.00 02.00 04.00 06.00 08.00 10.00 12.00 14.00 16.00 18.00 20.00 22.00
°C	37.50 37.00 36.30 36.20 36.30 36.50 37.25 36.90 37.25 37.30 37.35 37.40
FRIDAY	24.00 02.00 04.00 06.00 08.00 10.00 12.00 14.00 16.00 18.00 20.00 22.00 24.00
°C	37.50 37.00 36.50 36.20 36.30 36.40 36.30 37.20 37.25 37.24 37.25 37.40 37.60

Q2 Do you think Chris is an owl or a lark?

Q3 Mark a vertical band on your graph from 12.00 hours to 07.00 hours for Thursday and shade it lightly. Do the same for Friday. These are the hours Chris was sleeping.

Q4 How well do you think Chris would cope with a night shift from 21.00 hours to 04.00 hours (9.00 pm until 4.00 am). Why?

195

Temperature control

All endotherms have to prevent heat loss or encourage heat loss to regulate their body temperature.

Different endotherms solve these problems in different ways. It is the biologist's job—in this case your job—to explain how each endotherm warms up and cools down.

Collect
- simple equipment as required

1 Read through each of the problems below, choose one, and try to think of the most likely ways the endotherm is controlling its temperature.
2 Devise a simple experiment to test your idea. Your experiment should model the situation using simple equipment.
3 Check with your teacher that your design is workable and then carry out your experiment.

Problem 1
Many animals cluster and huddle together in cold weather like these penguins. Sometimes they change places with one another. Why do you think they do this?

Problem 2
Robins and other birds look different in summer than in winter. Can you decide which is which? What has this do to with temperature control?

Q1 Write a full report on your experiment.

Q2 Do your results support your idea?

Q3 Do you need to change your original idea?

Q4 How accurate and reliable was your experiment?

Q5 Did your experimental model over-simplify the process?

Osmosis

Here are some examples of osmosis that take place in your body every day.

Drinking a large amount of water

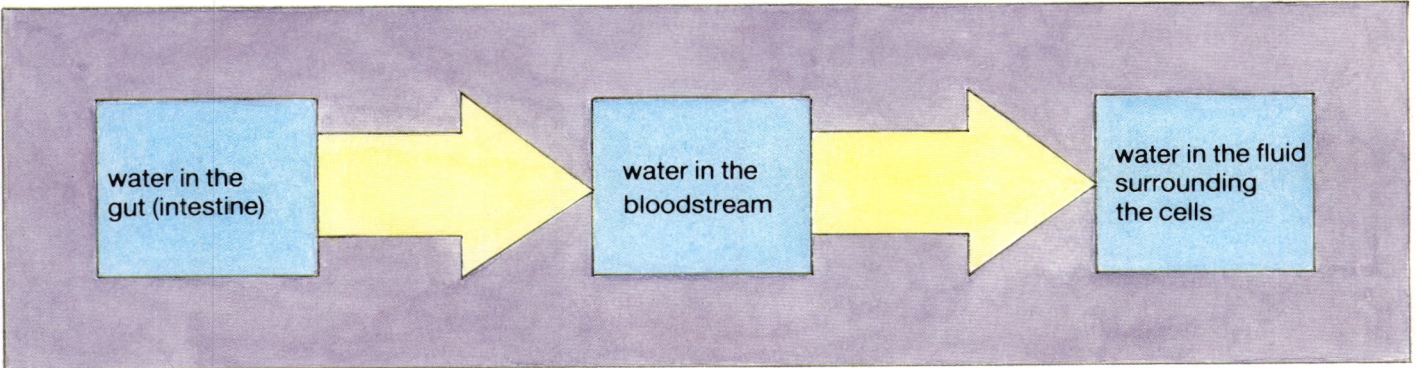

Q1 What happens to the following?
   a The amount of water in the cells.
   b The amount of water surrounding the cells.
   c The concentration of chemicals inside and outside the cells.

Eating salt (sodium chloride)

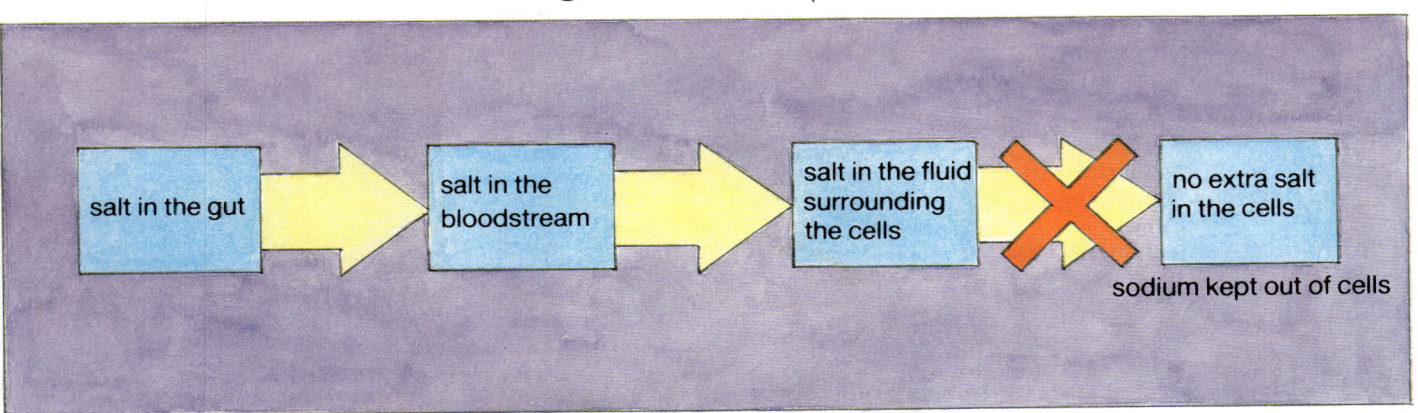

Q2 What happens to the following?
   a The amount of water in the cells.
   b The amount of water surrounding the cells.

Drinking water after eating salty food

Q3 How does the water balance inside and outside the cells change?

How are urea and water excreted by the kidney?

Go through the questions with a partner and make your own notes as you go along. Diagrams of different parts of the kidney will help your answers.

Q1 Look carefully at all the illustrations on this page and on page 140.

Q2 Examine the diagram of one complete human nephron. Talk about its shape with your partner. (Use words such as a **convolution** ('wriggly bit'), a **duct** (a tube or channel).)

Q3 Now look at the photographs of a small area of the cortex. Note the magnifications. Try and sort out how the photograph on the left shows nephrons.

Q4 Why do you think the medulla looks so different from the cortex?

Q5 Copy the diagram of the nephron and write in the names of the different parts.

Q6 What do you think is carried in the collecting duct? Where does it lead to?

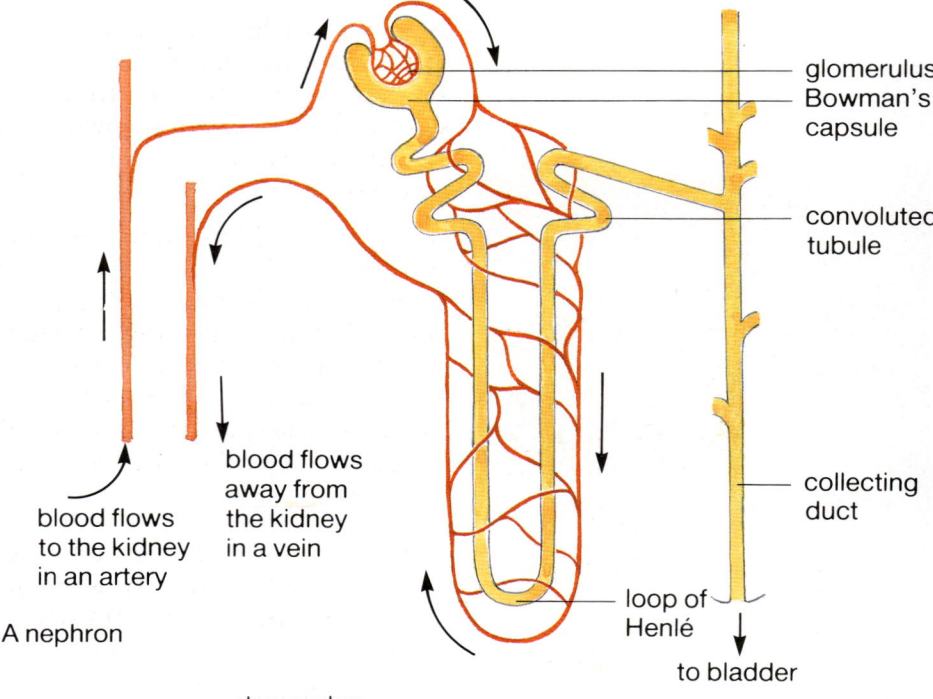

glomerulus
Bowman's capsule

convoluted tubule

collecting duct

loop of Henlé

to bladder

blood flows to the kidney in an artery

blood flows away from the kidney in a vein

A nephron

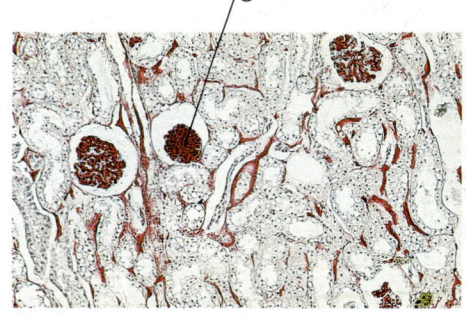

glomerulus

Glomeruli, ducts and blood vessels (×170)

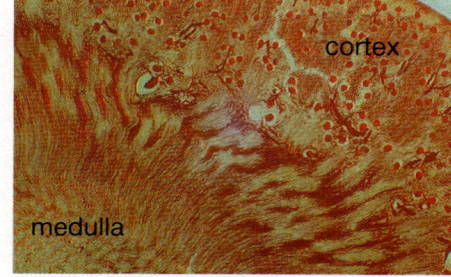

cortex

medulla

Section through cortex showing nephrons (×70)

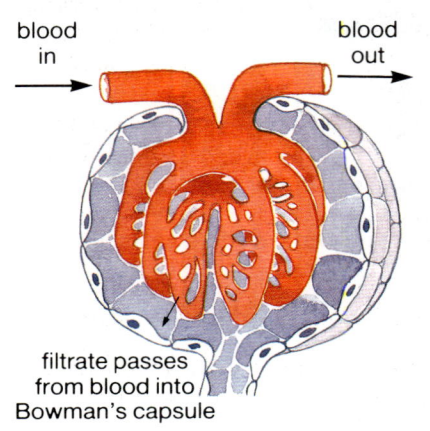

blood in

blood out

filtrate passes from blood into Bowman's capsule

Bowman's capsule and glomerulus

As blood is forced through a network of capillaries in the **glomerulus**, water with urea, glucose and dissolved salts like sodium (filtrate) filter into the **Bowman's capsule**, while blood cells and large molecules are held back. This process is called **ultra-filtration** (filtration under pressure).

As the filtrate passes through the first convoluted part of the tubule, glucose is reabsorbed into the blood stream. Water is also reabsorbed. This leads to an increase in the concentration of urea in the tubule. At the second convoluted part of the tubule more water and salts may be reabsorbed depending on the water balance of cells in the rest of the body and the amount of water and dissolved salts in the blood.

The remaining **urea** combined with water forms **urine**, which is stored in the **bladder** until it is excreted through the **ureter** which forms the **urethra** where the urine leaves the body.

Kidney failure

Although we have two kidneys it is possible to live healthily with just one. Sometimes both kidneys fail at the same time. This might be a temporary problem but often the kidneys have been damaged or are diseased and will not recover. Without both kidneys a person will die within two weeks because potassium salts build up in the blood and cause heart failure.

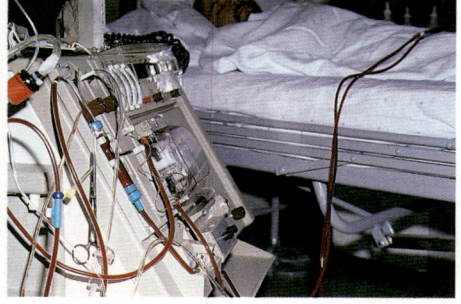

An artificial kidney does the same job as the real kidney by removing unwanted substances so that the blood has the correct concentration of salts and other molecules. It manages this in a different way from real kidneys by allowing poisonous substances to diffuse out of the blood. (Can you remember the name of the process by which real kidneys separate toxic waste from substances the body wants to keep?) The artifical membrane allows some particles through (such as ions and small molecules). This process is called **dialysis**.

Collect

- Visking tubing
- solution of protein, glucose and sodium chloride in distilled water
- beaker
- 3 test tubes
- flame loop
- silver nitrate solution
- safety glasses

1 Set up this experiment using Visking tubing. Remember to rinse the 'sausage' carefully before putting it in the beaker.
2 Leave for at least 30 minutes or until the next lesson.
3 Test the solution in the beaker to see what it contains. Use the following indicators:

- Benedict's reagent for glucose
- biuret reagent for protein
- flame test for sodium ions
- silver nitrate solution for chloride ions.

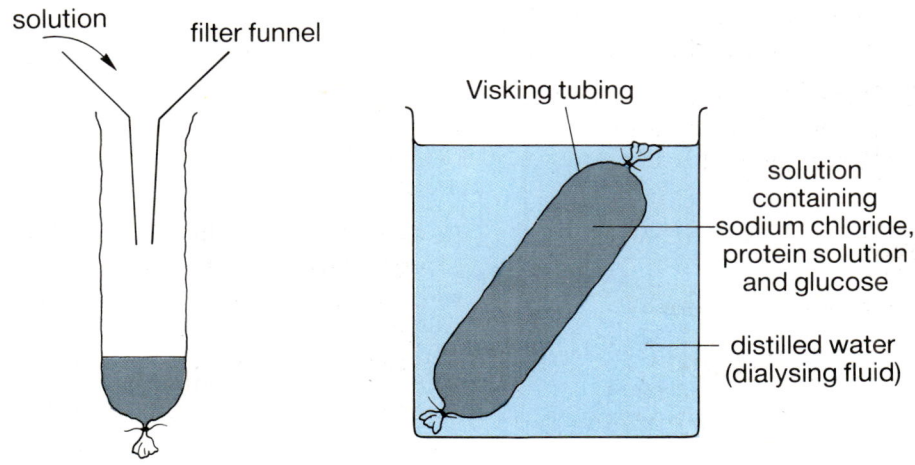

Q1 Write a report on your experiment. Explain all your results.

Q2 Remember that kidneys do not remove all the salt from blood. How could you change this experiment so that some of the sodium chloride but not all of it was dialysed out of the Visking tubing?

continued ▶

▶ continued

The artificial kidney has a large area of Visking membrane which is partially permeable. The patient's blood is pumped into the machine and flows along one side of the membrane. On the other side flows the **dialysing fluid**. The patient needs to lose some of the poisonous potassium salts from his/her blood but does not want to lose glucose and essential salts. To stop this happening the dialysing fluid has the same concentration of sodium chloride, sodium ethanoate, magnesium chloride, potassium chloride and calcium chloride as normal blood. Unwanted substances in the patient's blood like excess potassium chloride, excess sodium chloride and urea diffuse out into the dialysing fluid. Water is removed by partly closing the return tube (leading to the patient's vein) which raises the pressure of the blood entering the dialysis machine so that water is forced out through the Visking membrane. Sodium ethanoate diffuses into the blood and maintains the pH at its correct value.

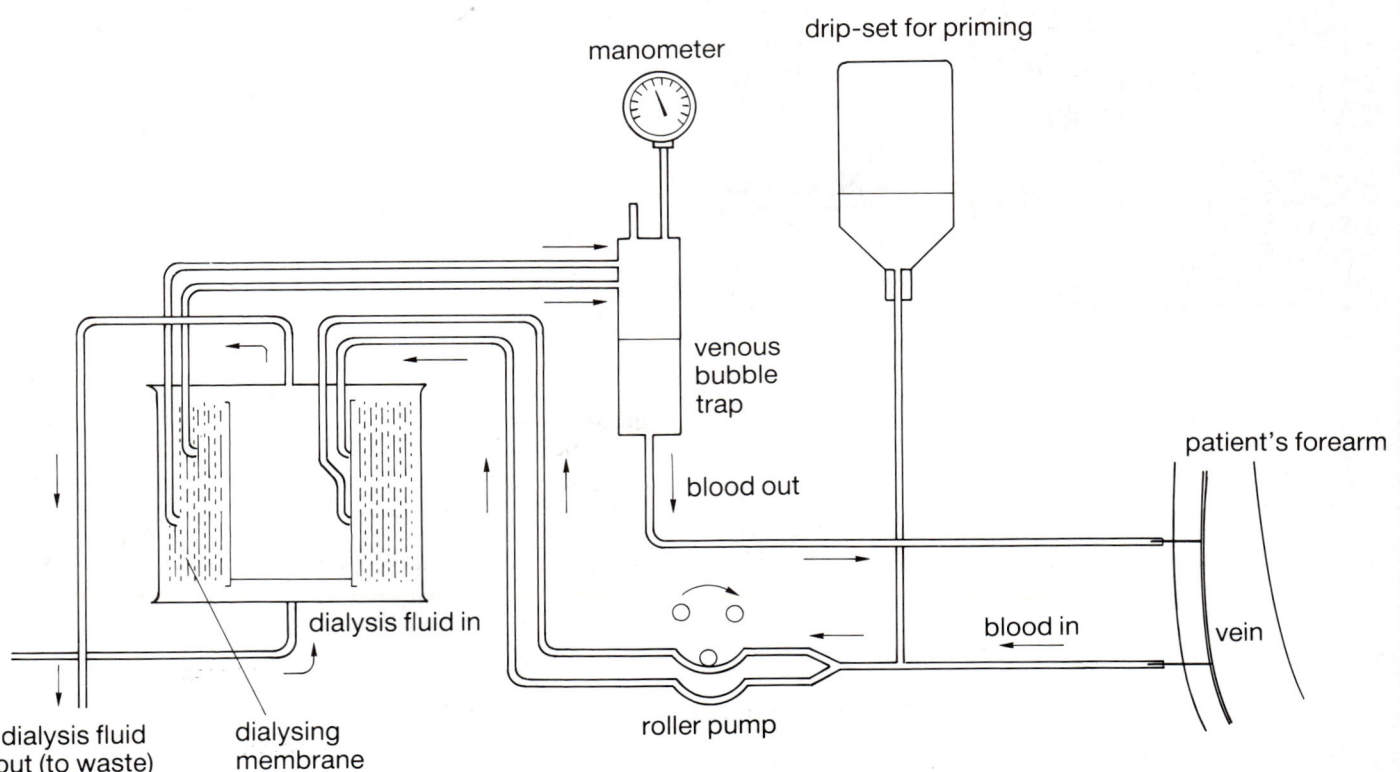

Dialysis in an artificial kidney circuit

Q1 Why is a large area of Visking membrane used?

Q2 Why is fresh dialysing fluid continuously pumped across the dialysis membrane?

Fact or fiction?

(above) Stanley Pons and Martin Fleischmann with their 'cold fusion' apparatus

FUSION CONFUSION

There has been a good deal of fuss recently about so-called cold fusion. Two scientists (one British) working in the US claim to have discovered an almost inexhaustible source of energy with apparatus no more complex than that found in school laboratories. This energy source would have none of the pollution problems of acid rain or the greenhouse effect associated with coal and oil fired power stations. Too good to be true? Well maybe, because so far other scientists have been unable to confirm these experimental results of Stanley Pons and Martin Fleischmann.

But what is fusion anyway? Fusion reactions occur in the sun and in thermonuclear bombs, but only at very high temperatures. These high temperatures make light atoms like hydrogen collide so fast that the centres of their atoms (their nuclei) squash together and merge. This is rather like squashing together two lumps of plasticine. This means that a different element is formed. In the sun, four atoms of hydrogen merge or fuse together to form an atom of helium.

Hydrogen atoms	→	Helium atom
4H		He

The helium atom has less mass than the four atoms of hydrogen and the difference is given out as heat energy. The energy output is so great that one *gram* of hydrogen would give out as much energy as the burning of 114 *tonnes* of petrol! Fleischmann and Pons believe that they have brought about fusion in a type of electrolysis experiment where they pass electricity through deuterium oxide – a type of heavy water. *Science Watch* will keep you informed of developments. ∎

Q1 What is fusion?

Q2 What is the fusion reaction that takes place in the sun?

Q3 What would the advantages of cold fusion be if the claim were true?

Q4 Why would people want the claim to be true? Do you think this might have influenced the belief of the media when it was first brought to their attention?

Q5 Are there any groups who would prefer the story not to be true?

Q6 What do you think those who care about the environment would think of the idea of cold fusion?

The mass spectrometer

The mass spectrometer is an instrument which is used for finding out the approximate mass of one atom compared with the mass of a hydrogen atom. This is called the **relative atomic mass** of an element. Because it is a comparison between two masses no unit is used.

Elements are mixtures of isotopes. Therefore the relative atomic mass of an element is an **average** of the mass of each type of isotope present in the sample.

Chlorine has two isotopes, ^{37}Cl and ^{35}Cl. In a sample of chlorine, 25% of atoms are ^{37}Cl and 75% are ^{35}Cl. There are three atoms of ^{35}Cl for every one atom of ^{37}Cl.

three have mass number 35	$3 \times 35 = 105$
one has mass number 37	$1 \times 37 = 37$
total mass of four atoms	$= 142$
average mass of four atoms	$= \dfrac{142}{4} = 35.5$

The relative atomic mass of chlorine is **35.5**

Key
1 The sample is vaporised.
2 The atoms in the vapour are bombarded by a stream of fast-moving electrons. The electrons collide with the atoms, knocking electrons off them and so producing positive ions.
3 An electric field speeds up the positive ions.
4 A magnetic field deflects the ions.
5 The deflected ions enter a detector which gives a line trace on a printout.

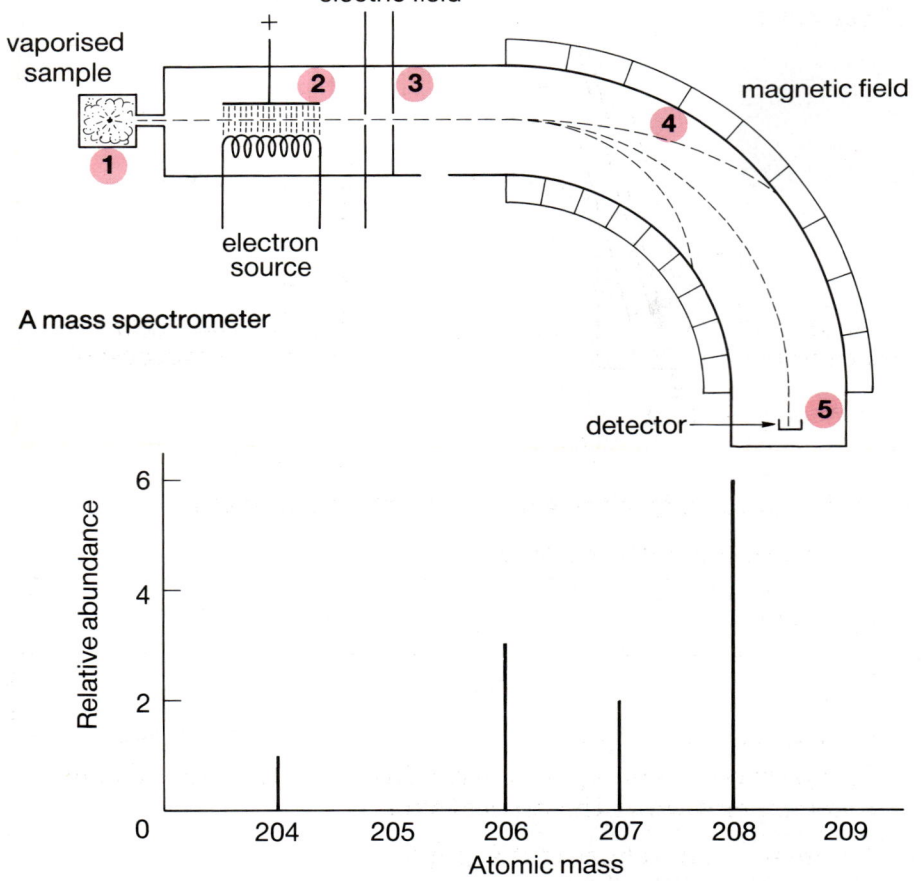

A mass spectrometer

The trace produced by the mass spectrometer shows that lead has four isotopes

Ions with the **same mass** are deflected by the same amount. When the magnetic field is changed ions with **different masses** are deflected giving a different line trace. In this way lines are drawn for all the different isotopes of an element sample.

Q1 What are the masses of the four isotopes of lead?

Q2 Which isotope is most abundant and which is least abundant?

Q3 Construct a table to show the numbers of protons, electrons and neutrons in each isotope of lead.

Q4 A mass spectrometer can be used to identify which elements are present in a substance. Explain how it does this.

Changing state of hexadecanol

You have learnt that covalent bonds between atoms in a molecule are very strong. However the bonds between the molecules are weak, as we found with graphite.

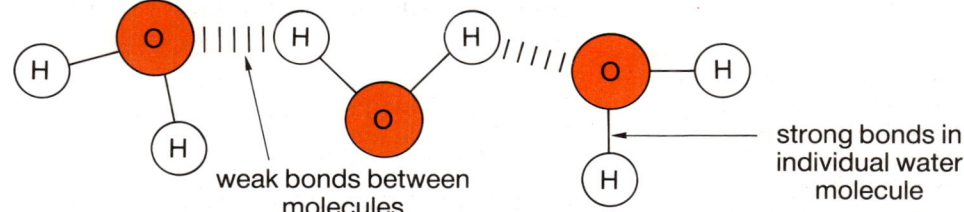

Water is an example of weak bonding between molecules

When a covalent substance melts, the weak bonds between the molecules break and the molecules separate. Energy is needed to break the bonds. The energy is provided in the form of heat. Conversely, when bonds are made, energy is given out.

Collect

- beaker
- boiling tube of hexadecanol
- bunsen burner
- heat-proof mat
- tripod and gauze
- thermometer
- clamp and stand
- stirrer
- safety glasses

1 Heat the water in the beaker and then stand the boiling tube in the beaker until the solid melts.
2 Turn off the bunsen. Put the boiling tube in the clamp.
3 Stir the liquid gently as it cools. Record the temperature every 30 seconds.

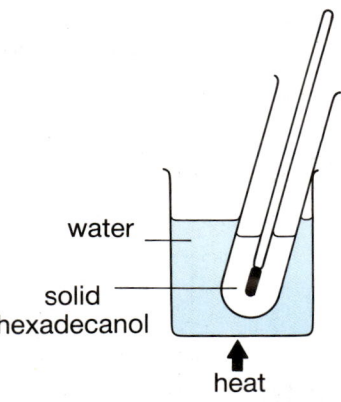

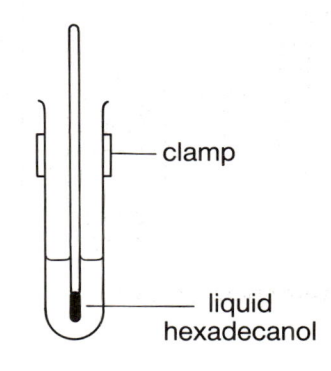

Q1 Write an account of the way you did the experiment.

Q2 Record your results in a table.

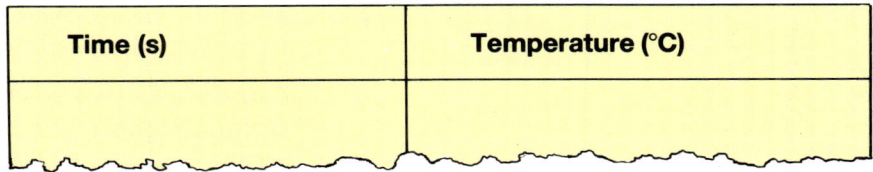

Time (s)	Temperature (°C)

Q3 Plot your results on a graph, with time on the horizontal axis and temperature on the vertical axis.

Q4 Comment on the shape of the graph.

Q5 What is the melting point of the hexadecanol?

Electrolysis of ionic solutions

When solutions of ionic compounds are electrolysed, water molecules are split up into hydrogen, produced at the cathode, and oxygen at the anode. (You should remember the process of electrolysis. Look up the details if you've forgotten!)

Collect

- the apparatus in the diagram
- potassium bromide solution
- potassium iodide solution
- copper sulphate solution
- safety glasses

1 Assemble the apparatus as shown.
2 Switch on the electricity.
3 Collect and test any gases formed at the electrodes.
4 Note any other observations.

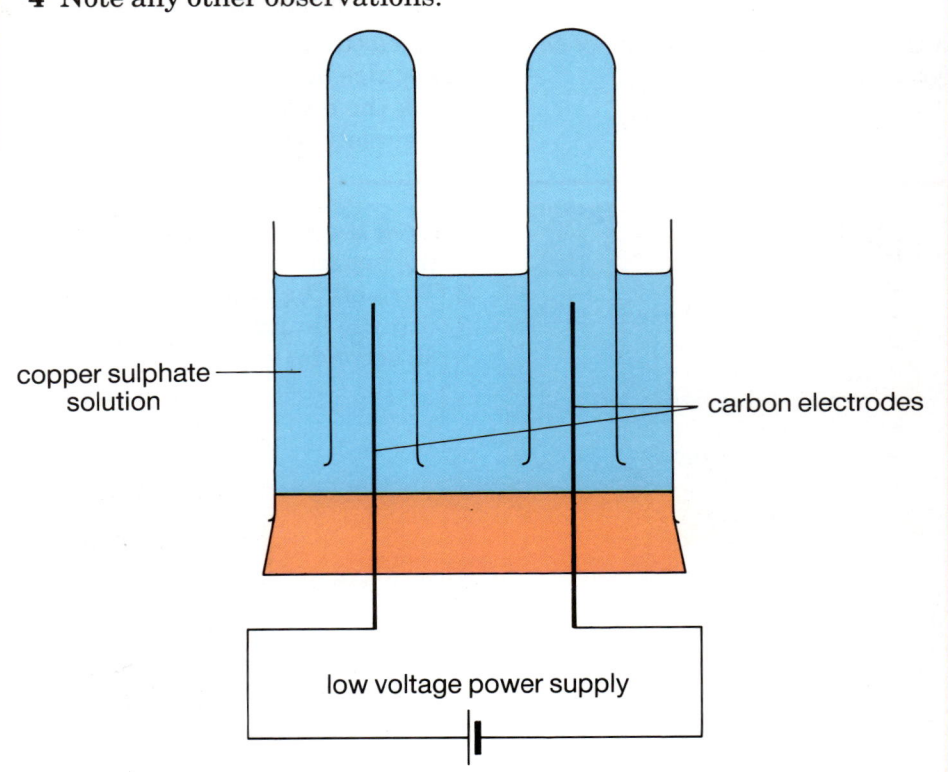

copper sulphate solution

carbon electrodes

low voltage power supply

Q1 Describe your experiment with the aid of a labelled diagram.

Q2 Record your observations in a table.

Q3 Copy and complete the following rule by filling in the gaps for electrolysis when either carbon or platinum electrodes are used.

Anode (positive electrode)
If the non-metal ion present in the compound is a **halide** (group . . . .) then a **halogen** is formed. If not then . . . . gas is formed.

Cathode
If the metal in the compound is low in the reactivity series then the metal is deposited. If the metal is high in the reactivity series then . . . . is formed.

Interpreting physical properties

Substance	Melting point (°C)	Boiling point (°C)	Electrical conductivity		
			solid	liquid	solution in water
A	1084	2570	yes	yes	insoluble
B	−218	−183	no	no	insoluble
C	−7	59	no	no	insoluble
D	801	1413	no	yes	yes
E	1610	2230	no	no	insoluble
F	660	2470	yes	yes	insoluble
G	−60	−20	no	no	yes

For each question give the letter or letters and explain why you chose it. Each question may have more than one answer.

Q1 Which substance is a metal?

Q2 Which substance consists of a giant lattice of ions?

Q3 Which substance is a gas at room temperature and pressure?

Q4 Which substance is a liquid at room temperature and pressure?

Q5 Which substance contains simple molecules?

Q6 Which substance is a macromolecule?

Q7 Which substance contains moving ions when molten or dissolved in water?

Q8 Which substance consists of molecules when pure but produces ions when dissolved in water?

Index

O

ocean ecosystems 18
ohm(s) 68
Ohm's law 68
oil 106
  cracking 91
optical fibres 33
optical instruments 176
ores 43, 44
organisms, where they live 10
osmosis 137, 197
oxidation reactions 93
oxygen
  atoms of 151, 152, 158, 159
  inspiration 135
  molecules of 159
  in water, content of 19
  in water molecule 158, 159

P

pancreas, insulin produced by 142
paraffin, heat output from 94
Peak District, mining in 42–3
Periodic Table 150, 153, 169
periscope 29
peroxidase 59
pH, of acids 46, 47
pH meter, neutralisation investigated
  with 53
photons 154
physical properties of substances,
  interpreting 205
phytoplankton 20
planets, gravitational field 110–11
plankton 20
polarisation of light 177
populations 14–17
  growth 15–17
potential difference (p.d.) 68, 78, *see also*
  volts
  measurement 67
power
  definition 70, 118
  of electrical appliances/devices
    70–71, 72
  output 118–19
power lines/cables 80–81, 185
precipitation
  of bases 50
  of insoluble salts 56
prisms 32
producers 10
propane, burning 92
protein, urine test for 145
protons 66, 154
  number of 155, 162
pulley, measuring efficiency of 120
pulsars 190

Q

quadrats 11
quasars 190

R

radiation
  electromagnetic 34–7
  of heat 130

radio(s) 105
  building 37
radio telescopes 190
radio waves 35
  from outer space 190
radioactive isotopes 156
reactions *see* chemical reactions
red blood cells, osmosis with 137
reflection of light 26, 32–3
  total internal 32–3
refraction of light 30, 33
relays 186
resistance, electrical 66, 68–9, *see also*
  ohms; Ohm's law
resistors
  in parallel 68, 69
  in series 69
respiration 135, 139

S

safety, car 123
salinity of seas 18
salts 54–7, *see also* sodium chloride
  insoluble 56
  soluble 54–6
shadows 175
shivering 132
silver atom 151
slaked lime 52
sodium, atoms of 151, 152, 162
sodium chloride (salt)
  crystals 163
  eating 197
sodium ethanoate 187
solar power 107
solids, particle arrangement in 134
soluble salts 54–6
solutions 187
  saturated and super-saturated 187
solvents 18
species, statistical sampling 11
specific heat capacity of water 194
spectrometer, mass 202
speed, rate of change of (=acceleration), force
  and mass related to 115, 122
stability of objects 115–16
static electricity 66, 183
stomach, hydrochloric acid in, too much 180
stretching things 113
sugar *see* glucose
sulphuric acid
  manufacture 48–9
  uses 48
surface area in chemical reactions,
  increasing 61
sweat and sweating 133, 140

T

telescopes 176
  radio 190
television 105
temperature
  body 126–33, 195–6
    control/regulation 128–33, 143–4,
     146, 196
    core 126
    shell 126
  in chemical reactions, increasing 60

thermal decomposition 52
thermography 36, 126, 127
thermoregulation in humans and
  animals 128–33, 143–4, 146, 196
thyroid gland 143
tidal power station 107
tides 192
  spring and neap 192
transducers, energy 103–5
transformers 78–81
trawl fishing 20

U

ultrafiltration by kidney 198
ultraviolet radiation 36
urea 138, 141, 198
ureter 198
urethra 198
urine 140, 198
  testing, in diagnosing illness 144,
   145

V

vasoconstriction 132
vasodilation 133
volt(s) and voltages 66, *see also* potential
  difference
  reduction by transformers 78
voltmeters 67

W

waste products of humans/animals,
  excretion 138, 198
wasting energy 191
water
  of crystallisation 187
  drinking 197
  evaporation 130
  excretion/loss from body 138, 140,
   198
  life under 18–21
  power, electricity produced from 106
  production from burning gases 92
  specific heat capacity 194
water molecules
  osmosis 137
  structure 158–9
watts 70, 118, *see also* kilowatthours
wavelength of electromagnetic radiation 34
weight 110
  atomic 154–7
wind turbines 107
wiring, household 184
wood as fuel 95–6, 108
work 119

X

X-rays 36, 39

Y

yeast cell, electricity from 98

Z

zinc–carbon cells 82
zooplankton 20